People, Land and Water

People, Land and Water

Participatory Development Communication for Natural Resource Management

Edited by
Guy Bessette

Routledge
Taylor & Francis Group

LONDON AND NEW YORK

First published by 2006 Earthscan and the International Development Research Centre (IDRC)

2 Park Square, Milton Park, Abingdon, Oxon OX14 4RN
711 Third Avenue, New York, NY 10017, USA

Routledge is an imprint of the Taylor & Francis Group, an informa business

First issued in paperback 2016

ISBN: 978-1-84407-343-6 (hbk)
ISBN: 978-1-138-97812-6 (pbk)

Typesetting by JS Typesetting Ltd, Porthcawl, Mid Glamorgan
Cover design by Mike Fell

A catalogue record for this book is available from the British Library

Library of Congress Cataloging-in-Publication Data

People, land, and water : participatory development communication for
natural resource management / edited by Guy Bessette.
      p. cm.
  Includes bibliographical references.
  ISBN-13: 978-1-84407-344-3 (hardback)
  ISBN-10: 1-84407-344-0 (hardback)
  1. Natural resources–Management. 2. Communication in economic
development. I. Bessette, Guy, 1952–
  HC85.P46 2006
  333.701'4–dc22
                                                    2006002676

Contents

List of Figures, Tables and Boxes

Figures

Tables

Boxes

List of Contributors

Mario Acunzo is a communication for development officer at the Extension, Education and Communication Service of the United Nations Food and Agriculture Organization (FAO) in Rome.

Claude Adandedjan is a senior education fellow at the International Centre for Research in Agroforestry (ICRAF), Sahel programme.

Awa Adjibade is a sociologist currently based in Lomé, Togo.

Madeline Baguio Quiamco is an assistant professor at the College for Development Communication of the University of the Philippines, Los Baños.

Guy Bessette is a senior programme specialist at the International Development Research Centre in Ottawa.

S.T. Kwame Boafo is Chief, Executive Office, Communication and Information Sector of the United Nations Educational, Scientific and Cultural Organization (UNESCO).

Maria Celeste Cadiz is Dean and Associate Professor at the College of Development Communication of the University of the Philippines, Los Baños.

Rawya El Dabi works for the International Development Research Centre's Partnership and Business Development Division in Cairo.

N'Golo Diarra is a researcher and trainer at the Centre de Services de Production Audiovisuelle (CESPA) in Mali.

Corinne Dick is at the American University of Beirut in Lebanon.

Vicenta P. de Guzman is the Executive Director of the Legal Assistance Centre for Indigenous Filipinos (PANLIPO) in the Philippines.

Waad El Hadidy is a programme manager at the Centre for Development Services in Egypt.

Mona Haidar is a researcher at the Environment and Sustainable Development Unit at the American University of Beirut in Lebanon.

Shadi Hamadeh is a professor of animal sciences and is currently leading the Environment and Sustainable Development Unit at the American University of Beirut in Lebanon.

Amri Jahi is a senior lecturer and researcher at Bogor Agricultural University in Indonesia.

Meya Kalindekafe is a senior lecturer in ecology at the University of Malawi.

Chris Kamlongera is the Director of the SADC Centre of Communication for Development.

Jones Kaumba is a senior communication for development trainer at the SADC Centre of Communication for Development.

Lun Kimhy is the Deputy Provincial Programme Adviser of the Partnership for Local Governance in Ratanakiri, Cambodia.

Yacouba Konaté passed away in 2003. At the time of his death, he was the coordinator of a participatory development communication project at the Permanent Inter-State Committee for Drought Control in the Sahel (CILSS).

Kofi Larweh is the Station Coordinator of Radio Ada, in Ghana.

Lourdes Margarita A. Caballero is a research associate at the College of Development Communication of the University of the Philippines, Los Baños.

Luningning A. Matulac is a professor of educational communication at the College of Development Communication of the University of the Philippines, Los Baños.

Pierre Mumbu is a researcher and lecturer at the Institut Supérieur de Développement Rural of Bukavu, in the Democratic Republic of Congo and also works as a consultant in community radio and development communication.

Amadou Niang is the Director of the Millennium Development Goals Centre for West Africa, in Bamako.

Michelle Obeid is a researcher at the Environment and Sustainable Development Unit at the American University of Beirut in Lebanon.

Nora Naiboka Odoi is a development communication specialist at the Kawanda Agricultural Institute in Uganda.

Kadiatou Ouattara is a journalist and works with Journalistes en Afrique pour le Développement (JADE) in Burkina Faso.

Souleymane Ouattara is a journalist and the Coordinator of Journalistes en Afrique pour le Développement (JADE) in Burkina Faso.

Rosalie Ouoba is a sociologist, based in Burkina Faso. She is actively involved with the Union of Rural Women of West Africa and Chad.

Sours Pinreak is an adviser to the Land Rights Extension team in Cambodia.

Nora Cruz Quebral is recognized as the founder of the discipline of development communication. She currently heads the Nora Cruz Quebral Foundation for Development Communication and is still associated with the College of Development Communication of the University of the Philippines, Los Baños.

C. V. Rajasunderam, now retired, has worked as a consultant in development communication.

Chin Saik Yoon is the Publisher and Managing Director of Southbound in Malaysia.

Ahmadou Sankaré is the coordinator of a participatory development communication project at the Permanent Inter-State Committee for Drought Control in the Sahel (CILSS) in Burkina Faso.

Karidia Sanon is a researcher and lecturer at the Université de Ouagadougou, Burkina Faso.

Fatoumata Sow is a journalist currently working at UNESCO's regional office in Senegal.

Diabodo Jacques Thiamobiga is an agronomist and sociologist working in the field of rural development in Burkina Faso.

Jakob S. Thompson is an associate professional officer at the United Nations Food and Agriculture Organization (FAO) in Rome.

Cleofe S. Torres is an associate professor at the Centre for Development Communication of the University of the Philippines, Los Baños.

Le Van An heads the Department of Science and International Relations at the Hue University of Agriculture and Forestry.

Maria Theresa H. Velasco is an associate professor and Chair of the Department of Science Communication at the College of Development Communication of the University of the Philippines, Los Baños.

Rami Zurayk is a researcher at the Environment and Sustainable Development Unit at the Ameircan University of Beirut, in Lebanon.

Foreword

Nora Cruz Quebral

For many communication professionals who have made development a personal commitment, shuttling from theory to practice and then from application to re-conceptualization is what their calling is about. For this breed, the 'field' is the main laboratory and proving ground. Working there can be frustrating, uplifting and chastening, but never dull, as the contributors to this book can attest, particularly so when one is questioning seminal concepts and trying out new ones in their place, which is what participatory development communication (PDC) does in this book.

Development communicators, being new kids on both the development and communication blocks, may not quite have gained full acceptance among their social science peers. They are a dynamic lot, nonetheless, who stake their alternate positions with passion or with studious persistence. A sign of their independence is the several names that they have given their specialty. In addition to PDC, this book cites participatory communication for development, participatory communication or communication for social change. Participation and dialogue are givens in all the variants.

There is none better systematized than PDC, however. This book offers a methodology and a terminology honed in the field. Based on a succession of 'writeshops', workshops and round-table conferences attended by action researchers and outreach workers from Asia and Africa, the collective experience was in itself an experiment in cross-continental dialogue. The linguistic and cultural differences were not trivial. They were a macrocosm of the divides that practitioners and field researchers normally encounter in their local communities.

The group effort recounts diverse interventions in which development communication concepts were tested, modified or found to be a fit. The overall results should encourage other fieldworkers to enlarge the discussion further with their own experiments. For the contributors to the book, re-reading the cases at leisure will still be part of their learning experience. They can better appreciate the conceptual differences between communication approaches and match up an approach with the appropriate methods and terms.

The friendships and increased understanding of each other's ways that ensued among the contributors were equally valuable outcomes of their PDC adventure. They bode well for a global PDC network, which is probably what the organizers had in mind all along.

As an endnote, the words 'participatory development communication' – singly or joined – connote certain values that define PDC as much as its strategies, tools and techniques. They are what makes the term different from other types of communication and are at the core of its guiding philosophy. PDC professionals are reminded to let those values shape their practice and to remember, as well, that sustainable natural resource management is but a facet of the larger goal that is human development.

Manila
April 2006

Preface

This book presents conceptual and methodological issues related to the use of communication in order to facilitate participation among stakeholders in natural resource management (NRM) initiatives. It also presents a collection of chapters that focus on participatory development communication and NRM, particularly in Asia and Africa.

There are many approaches and practices in development communication, and most of them have been implemented in the field of environment and natural resource management. But, even when considering participatory approaches in NRM, communication is often limited to information dissemination activities that mainly use printed materials, radio programmes and educational videos to send messages, explain technologies or illustrate activities. These approaches, with their strengths and weaknesses, have been well documented.

Participatory development communication takes another perspective. This form of communication facilitates participation in a development initiative identified and selected by a community, with or without the external assistance of other stakeholders. The terminology has been used in the past by a number of scholars[1] to stress the participatory approach of communication in contrast with its more traditional diffusion approach. Others refer to similar approaches as participatory communication for development, participatory communication or communication for social change.

In this publication, participatory development communication is considered to be a planned activity that is based on participatory processes and on media and interpersonal communication. This communication facilitates dialogue among different stakeholders around a common development problem or goal. The objective is to develop and implement a set of activities that contribute to a solution to the problem or the realization of a goal, and which support and accompany this initiative.[2]

This kind of communication requires moving from a focus on information and persuasion to facilitating exchanges between different stakeholders to address a common problem, to develop a concrete initiative for experimenting with possible solutions, and to identify the partnerships, knowledge and materials needed to support these solutions.

This book situates the concept and its methodological issues. It has been produced through a three-step process. First, practitioners from Asia and Africa were invited to submit chapters that offer examples and illustrations of applying participatory development communication to natural resource management. Second, a peer-review workshop was organized in Perugia, Italy, in September 2004, in preparation for the Roundtable on Development

Communication organized by the United Nations Food and Agriculture Organization (FAO) to discuss and review these chapters. Third, during the roundtable, the first chapter of this volume was presented and introduced to the participants in order to orient the discussions of the working group on communication and natural resource management.

These steps led to the preparation of this book, which we hope will play a role in both promoting participatory approaches to development communication in the field of environment and NRM, and in sharing the viewpoints of practitioners from Asia and Africa.

Guy Bessette
April 2006

Notes

1   See, in particular, White et al (1994) and Servaes et al (1996).
2   See Bessette (2004).

References

Bessette, G. (2004) *Involving the Community: A Facilitator's Guide to Participatory Development Communication*, IDRC, Ottawa, Canada, and Southbound, Penang

Servaes, J., Jacobson, T. L. and White, S. A. (1996) *Participatory Communication and Social Change*, Sage Publications, London

White, S. A., Sadanandan Nair, K. and Ascroft, J. (1994) *Participatory Communication, Working for Change and Development*, Thousand Oaks, New Delhi and Sage Publications, London

Acknowledgements

I would like to thank the contributors, who were willing to share their experiences and reflections with regard to participatory development communication and natural resource management.

My thanks also go to Manon Hogue, who very patiently and with a lot of enthusiasm revised the different chapters and accompanied them with a note of introduction. Without her work and support, there is no way this volume could have been finalized for publication. Her commitment and skills brought to this book much more than a technical contribution. I and the other authors of this publication salute you and thank you, Manon.

List of Acronyms and Abbreviations

AIDS	acquired immune deficiency syndrome
ARDA	Association for Rural Development in Arsaal (Lebanon)
BACDI	Bayagong Association for Community Development Inc (the Philippines)
CBCRM	community-based coastal resource management
CBNRM	community-based natural resource management
CCD	Convention to Combat Desertification
CDC	College of Development Communication (University of the Philippines)
CDS	Centre for Development Studies
CEDRES	Centre d'Études pour le Développement Économique et Social (Burkina Faso)
CESAO	Centre d'Études Économiques et Sociales de l'Afrique de l'Ouest
CESPA	Centre de Services de Production Audiovisuelle (Mali)
CIERRO	Centre for Rural Radio Development (Burkina Faso)
CILSS	Permanent Interstate Committee on Drought Control in the Sahel
ComDev	classical communication for development
DAES	Department of Agricultural Extension Services (Malawi)
DRC	Democratic Republic of Congo
FAO	United Nations Food and Agriculture Organization
GIS	geographical information system
GUCRE	Gestion des Usages Conflictuels des Ressources en Eau (Management of Conflicting Uses of Water Resources) project
ha	hectare
HIV	human immunodeficiency virus
IB-ESA	Isang Bagsak East and Southern Africa
ICRAF	International Centre for Research in Agroforestry
ICT	information and communication technology
IDRC	International Development Research Centre
IKS	indigenous knowledge system
IPD	Institut Panafricain pour le Développement (Burkina Faso)
JADE	Journalistes en Afrique pour le Développement (Journalists in Africa for Development)
KARI	Kawanda Agricultural Research Institute (Uganda)
km	kilometre
KVIP	Kumasi ventilated improved pit toilet
m	metre

mm	millimetre
NAADS	Integration of Natural Resource Management in National Agricultural Advisory Services (Uganda)
NARO	National Agricultural Research Organization (Uganda)
NEF	Near-East Foundation
NEPAD	New Partnership for Africa's Development
NGO	non-governmental organization
NRM	natural resource management
PANLIPI	Legal Assistance Centre for Indigenous Filipinos
PDC	participatory development communication
PLA	participatory learning and action
PLG	Partnership for Local Governance (Cambodia)
PNKB	Kahuzi-Biega National Park
PRA	participatory rapid appraisal
PRA	participatory reflection and analysis
PRA	participatory rural appraisal
PRCA	participatory rural communication appraisal
RIA–3	Research Institute for Aquaculture – Region 3
SADC-CCD	Southern Africa Development Community Centre of Communication for Development (Zimbabwe)
UFROAT	Union of Rural Women of West Africa and Chad
UK	United Kingdom
UNDP	United Nations Development Programme
UNESCO	United Nations Educational, Scientific and Cultural Organization
UNICEF	United Nations Children's Fund
US	United States

Introduction

I

Facilitating Dialogue, Learning and Participation in Natural Resource Management

Guy Bessette

Poverty alleviation, food security and environmental sustainability: The contribution of participatory communication

Poverty alleviation, food security and environmental sustainability are closely linked and represent major development challenges for all actors involved in the field of natural resource management. Poverty alleviation requires sustained economic growth, but it is also necessary to ensure that the poor benefit from that growth. Efforts must also be made to improve food security, not only through an increase in productivity, but also by providing conditions of access and proper utilization by the poor.

Promoting environmental sustainability includes challenging goals such as fighting land degradation (especially desertification), halting deforestation, promoting proper management of water resources through irrigation schemes and protecting biodiversity. All these activities must be designed and implemented with the active participation of those families and communities who are struggling to ensure their livelihood in changing and unfavourable environments. But they must also include other stakeholders who are playing or can play a role in these changes: government technical services, non-governmental organizations (NGOs), development projects, rural media, community organizations and research teams. Finally, local and national authorities, policy-makers and service providers must also be involved in shaping the regulatory environment in which the required changes will take place.

Effectively addressing the three interlinked development challenges of poverty alleviation, food security and environmental sustainability requires that development practitioners work actively with all stakeholders with a view to facilitating dialogue, learning and active participation in natural resource management initiatives.

Best practices in natural resource management research and development point to situations in which community members, research or development team members and other stakeholders jointly identify research or development parameters and participate in the decision-making process. This process goes beyond community consultation and participation in activities identified by researchers or programme managers. In best case scenarios, the research or development process itself generates a situation of empowerment in which participants transform their view of reality and are able to take effective action.

Participatory development communication reinforces this process. It empowers local communities to discuss and address natural resource management practices and problems and to engage other stakeholders in building an improved policy environment.

But what about the issues involved in applying participatory development communication to natural resource management practices and research? What are the challenges and the difficulties associated with this approach? What insights and lessons can be drawn from our practices in the field? This chapter offers a reflection on these issues and suggests orientations to further reinforce natural resource management practices and research through participation and communication.

Moving from information dissemination towards community participation

Traditionally, in the context of environment and natural resources management, many communication efforts used to focus on the dissemination of technical packages towards the end-users who were expected to adopt them. Researchers wanted to 'push' their products to communities and development practitioners in order to receive community commitment to their development initiatives.

Not only did these practices have little impact, but they also ignored the need to address conflicts or policies.

Participatory development communication takes a different approach. It suggests a shift in focus from informing people with a view to changing their behaviours or attitudes to facilitating exchanges between various stakeholders. These exchanges help the stakeholders to address a common problem or implement a joint development initiative in order to experiment with various solutions and identify the required partnerships, knowledge and material conditions.

The focus is not on information to be disseminated by experts to end-users. Rather, it is on horizontal communication processes that enable local communities to identify their development needs and the specific actions that could help to fulfil those needs, while establishing an ongoing dialogue with the other stakeholders involved (e.g. extension workers, researchers and decision-makers). The main objective is to ensure that the end-users gather enough information and knowledge to carry out their own development initiatives, evaluate their actions and recognize the resulting benefits.

Such a communication process pursues objectives related to increasing the community knowledge base (both indigenous and modern); modifying or reinforcing common practices related to water use and soil productivity so that natural resources can be managed more efficiently; building and reinforcing community assets; and approaching local and national authorities, policy-makers and service providers. Appropriate communication approaches should also be set up to implement the required initiatives, as well as to monitor and evaluate their impact, while planning for future action.

With participatory development communication, researchers and prac-titioners become facilitators in a process that involves local communities and other stakeholders in the resolution of a problem or the achievement of a common goal. This, of course, requires a change in attitude. Learning to act as a facilitator does not happen overnight. One must learn to listen to people, to help them express their views and to assist them in building consensus for action. For many natural resource management researchers and practitioners, this is a new role for which they have not been prepared. How can they initiate the process of using communication to facilitate participation and the sharing of knowledge?

Some of the chapters presented here describe this process in action. In Part III, in 'From Rio to the Sahel: Combating Desertification', Sankaré and Konaté describe how such an approach was developed in the context of desertification. Communication strategies were used to stress information dissemination, mobilization and persuasion; but they had little impact. There was a need to try out and implement other approaches. An experiment in participatory communication was used to support various local initiatives designed to fight desertification in the Sahel and to facilitate community participation.

The process led community members and local development actors to identify the problems facing them with regard to desertification, to express their needs and to decide on local solutions and concrete initiatives to experiment with. The project used communication tools such as practical demonstrations, and radio and community discussions, as well as traditional songs and poems to support and accompany the initiatives.

The process included four main phases: training, planning, experimentation and evaluation. Training and planning were the foundation because they mobilized all actors (e.g. community members, project leaders and communication facilitators from the locality) to discuss the process of the action research and how communication would be used to facilitate participation. Not only did this process facilitate community participation, but it also contributed to creating synergy between various development structures.

These initiatives were successful because people were involved in the decision-making process and were not simply invited to participate in specific activities. The project also demonstrated that halting desertification, like other development challenges, demands community participation and synergy between different development actors. It cannot be programmed in a top-down way.

In 'Growing Bananas in Uganda: Reaping the Fruit of Participatory Development Communication' Odoi tells the story of how the shift was made to implement communication for participation in the context of action research with banana growers. The banana research programme of Uganda's National Agricultural Research Organization (NARO) wanted to develop a two-way communication strategy to enhance farmer participation in experiments with different technologies in order to improve banana production and foster farmer-to-farmer training using communication tools developed in a participatory manner. This research used participatory development communication as a tool to foster the active participation of the community in identifying and solving their natural resource management problems.

Researchers encouraged farmers to form farmers' groups. They then helped the farmers' groups' representatives to identify and prioritize their natural resource management problems within their banana gardens, as well as to find the causes and potential solutions to these problems. The researchers also worked with the farmers to identify their communication needs and objectives regarding the identified problems, the activities that could be undertaken to alleviate the problems and the communication tools that could assist the farmers in sharing their new knowledge with their farmers' groups.

During this process, the researchers discovered that some farmers already had the appropriate knowledge concerning the natural resource management concerns that were raised, but that this knowledge could be reinforced. They also noted that farmers did not have a forum within which they could share information with each other. Thus, there was a need for communication tools.

As a result of the research activities, plots of land that farmers had previously abandoned started yielding good bananas. Farmers also grew confident enough to show their plots to other farmers and to share their knowledge with other farmers in their community. They learned to use communication tools such as photographs, posters, brochures, songs and dances.

After they appreciated the power of belonging to a group, they created a formal farmers' association through which they could search, access and share relevant information and services about community problems. As a result of these activities, the farmers have become proactive instead of passively waiting for external assistance.

A research action project in the basin of the Nakanbe River in Burkina Faso (see 'Water: A Source of Conflict, a Source of Cohesion in Burkina Faso' by Sanon and Ouattara) is another example of a participatory communication approach that brought all the stakeholders together to manage community conflicts related to water.

Approaches to water resource management are often centralized and allow for little participation of the local populations that are actually affected by water issues. Field research conducted in this basin revealed that 50 per cent of modern water sources (hand pumps and modern wells) that had been established by different projects were non-functional as a result of the lack of involvement and ownership by the beneficiaries. The participatory communication approach used by the research team favoured the use of two-way communication and emphasized dialogue among the different stakeholders with regard to water use. The approach also focused on capacity-building at the local level in terms of organization, participation and decision-making in water resource management and conflict resolution, and in implementing or reinforcing local water management committees.

Once again, participatory communication was helpful in identifying solutions to conflict situations in the villages and in setting up or reinforcing social institutions, such as the water management committee. However, it also built community members' confidence in their capacity to address their problems and to seek their own solutions, rather than wait for external assistance. In this case, the central role played by women in the management of water resources in the villages was also recognized.

Another case from Viet Nam (see 'Engaging the Most Disadvantaged Groups in Local Development: A Case from Viet Nam' by Le Van An) describes how a participatory communication approach was used to reinforce community-based natural resource management (CBNRM) research with upland communities. The research started after new policies were put in place by the government to protect forests in the uplands. However, following these measures, only 1 per cent of the land was left available for agricultural production. Thus, local communities who used to practise swidden agriculture had to change their practices and move to sedentary farming. This research initiative tried to help them improve their livelihood in this new context.

Due to these forced changes in their farming system and low access to assets and natural resources, production was low and there were few opportunities for income generation. Participatory communication was used to foster the participation of these local communities in identifying their needs and priorities, and to discuss ways of improving their livelihood. This approach introduced a change from the traditional ways of intervening in a given community from the outside. For the first time, groups of farmers who shared common characteristics and interests were asked what problems they wanted to start working on and what solutions they wanted to experiment with.

The question of reaching the poor and the most disadvantaged groups in the community was a major preoccupation because these people had few opportunities to participate in research or development programmes. Emphasis was put on the participation of poor farmers and of women. Improving the capacity of the commune's leaders and organizations also helped them to apply such participatory approaches with community members so that they could contribute to community plans and activities.

The natural resource management practitioner as a communication actor and facilitator

Establishing relationships

As soon as a researcher or natural resource management (NRM) practitioner first contacts a local community to establish a working relationship, that person becomes a communication actor. The way in which the researcher or NRM practitioner approaches the local community, understands and discusses the issues, and collects and shares the information involves methods of establishing communication with people. The way in which communication is established and nurtured affects how people feel involved in the issues and how they participate, or not, in the research or development initiatives at issue.

Within this framework, it seems important to promote a multidirectional communication process. The research team or the development workers approach the community through community leaders and community groups. The community groups define their relationship with the new resource people, with other stakeholders and with other community groups.

Many researchers still perceive community members as beneficiaries and future end-users of research results. Even if most people recognize that the one-way delivery of technologies to end-users simply has little impact, the shift in attitudes and practices is not easy. For this shift to happen, one must recognize that community members are stakeholders in the research and development process. Therefore, approaching a community also means involving people and thinking in terms of stakeholder participation in the different phases of the research or development process as a whole. Building mutual trust and understanding is a major challenge at this stage and will continue to be so during the entire period of interaction between researchers or practitioners and the community.

Negotiating mandate

Researchers do not come to a community without their own mandate and agenda. At the same time, communities also want their needs and problems to be addressed by resource people who approach them. Most of the time, they will not distinguish between NRM problems, difficulties in obtaining credit or health issues because these are all part of their reality.

Because they cannot address all these issues, researchers and practitioners should explain and discuss the scope and limitations of their mandate with community members. In some cases, compromises can be found. For example, it may be possible to involve other resource organizations that could help to resolve problems which are outside the mandate of the researchers or practitioners. This can often be the case with the issue of credit facilities.

Power relations and gender roles

The management of natural resources is clearly linked to the distribution of power in a community and to its socio-political environment. It is also closely associated with gender roles. This is why social and gender analyses are useful tools for examining the distribution of power in a community. Failure to use these tools may turn the participatory process into a manipulation process or make it selective of only a few individuals or groups in the community.

The chapter concerning communication and sustainable development in Part IV (see 'From Information to Communication in Burkina Faso: The Brave New World of Radio' by Ouattara and Ouattara) refers to a situation where a traditional healer had an unquestionable authority over everything that concerned the community, and used the participatory communication process to reinforce his authority over the community. The members of the intervention team, who were not used to such behaviour, were *de facto* manipulated by the situation. What kind of participation was then possible?

This situation is not exceptional and can only be prevented by identifying the main actors in a community and understanding their roles and relationships before any process is launched. Social analysis, gender analysis and identification of local communication systems, tools and channels should take place before any intervention that involves people in identifying problems and solutions.

Understanding the local setting: Collecting data or co-producing knowledge?

This attitude change has its corollary in methodology. Researchers have been trained in data collection, which emphasizes an extractive mode that does not facilitate participation. Participatory development communication (PDC), however, suggests that researchers or practitioners collaborate with community members and other stakeholders to assemble and share baseline information. This points to a process of co-producing knowledge that draws on the strengths of the different stakeholders.

Participatory rapid appraisal (PRA) and related techniques have been widely used in the field of natural resource management to assemble baseline information in record time with the participation of community members. However, we often find situations where techniques such as collective mapping of the area, transect walks, problem ranking and development of a timeline are still used in an extractive mode. The information is principally used for the researcher's or the project designer's benefits, and little consideration is given to the information needs of the community or to any restitution activity that would ensure the sharing of results.

In these cases, even with the 'participatory' label, these techniques can reinforce a process guided from the outside. PDC stresses the need to adapt attitudes as well as techniques. Co-producing knowledge differs from simply collecting data, and it can play an essential role in facilitating participation in the decision-making processes involved in a research or development project.

Understanding the communication context

Who are the different groups that comprise the local community? What are the main customs and beliefs regarding the management of land and water, and how do people communicate among themselves on these issues? What are the effective interpersonal channels of communication? What views are expressed by opinion leaders or exchanged by people in specific places? What local associations and institutions do people use to exchange information and points of views? What modern and traditional media does the community use?

Here again, we find value in integrating the biophysical, social and communication aspects within an integrated effort to understand the local setting. In the same way that they collect general information and conduct PRA activities to gather more specific information, researchers and development practitioners should seek to understand, with the help of the community, its communication channels, tools and global context.

Identifying and using local knowledge

Identifying the local knowledge associated with natural resource management practices is part of the process of co-producing knowledge. It should also be linked with two other issues: the validation of that knowledge and the identification of modern and scientific knowledge that could reinforce it.

Specific local knowledge or practices may be well suited to certain contexts. In other contexts, it may be incomplete or have little real value. Sometimes, specific practices may have been appropriate for previous conditions, but these conditions may have changed. This emphasizes the importance of validating common local knowledge against scientific evidence and through discussions with local experts or elders, as well as community members. It may also prove useful to combine and blend modern knowledge with local practices to render the latter more effective or more suited to local needs.

Another point is worth noting with regard to the use of local knowledge within a participatory communication approach. The process should not be conducted in an extractive mode by people outside the community. It should be a decision made by the community which is searching for some solutions to a given problem. Two chapters discuss issues related to participatory communication and local knowledge.

In the research conducted by Ouattara and Ouattara on communication and sustainable development in Part IV ('From Information to Communication in Burkina Faso: The Brave New World of Radio'), women from the community were trained as facilitators, and separate meetings were conducted with men and women. The facilitators always explained to the women the importance of their knowledge in the search for solutions to a specific problem.

A modern solution to a given problem will also have more chance of being adopted if a similar practice already exists in the community. For example, in the Sahel, the use of rocks to protect fields against erosion found easy acceptance because people already used dead branches to stop water from invading their fields.

In 'The Old Woman and the Martins: Participatory Communication and Local Knowledge in Mali' in Part III, Diarra reports on a case from Mali where ancient knowledge was used to improve agricultural production and the well-being of the community. An old woman in the village could predict good rain years and drought years and orient farmers to cultivate either on the high tablelands (during years of good rain) or by the side of the river (during years of drought). For this reason, each family had two plots of land, one by the riverside and the other one in the tablelands. Her well-protected secret was that she could make these predictions by observing the height at which martins built their nests in the trees near the river.

After her death, and with the permission of the village authorities, her story was told to the villagers in order to motivate the community to protect the shallow river from too much bank erosion. The villagers agreed to participate in such activities to protect the birds and the knowledge that they brought with them each year. This story tells us how local knowledge may be used in day-to-day lives and also motivate people to better manage their resources.

Involving the local community in diagnosis and planning

Participatory development communication also requires that the local community be involved in identifying a development problem (or a common goal), discovering its many dimensions, identifying potential solutions (or a set of actions) and making a decision concerning which ones to experiment with or implement. It also means facilitating interaction and collaborative action with other stakeholders who should be part of the process.

Traditionally, many researchers and practitioners used to identify a problem in a community and try out possible solutions with the collaboration of local people. With participatory development communication, the researcher or development practitioner becomes a facilitator of a process that involves local communities and other stakeholders in identifying and resolving a problem or achieving a common goal.

The communication process should help people to identify a specific problem that they want to address; discuss and understand the causes of the problem; and identify possible solutions and decide on a set of activities to experiment with. It is useful to stress that this does not happen during the course of a single meeting with community representatives. Time is needed for this process to mature.

In some cases, the departure point is not a specific problem but a common goal that a community sets for itself. As with the problem-oriented process, the community will decide on a set of actions to try to achieve that goal. At the end of both processes, the community decides on a concrete set of actions.

Ideally, development and research objectives should be identified at this point to strengthen and accompany the chosen community initiative. However, generally speaking, these objectives have already been identified in a research and development proposal that was conceived before the process was undertaken with the community. One solution to this problem is to plan a revision of the initial objectives with the community at the start of the research or development project. But, ideally, the administrative rules of donor organizations and the methodological habits of practitioners should be modified to facilitate community participation at the identification phase of what could become a research or development initiative.

Developing partnerships at the local level

The concept of developing partnerships between all development stakeholders involved with local communities is central to participatory development communication. We often find situations where a research or development initiative is conducted with a local community without considering other initiatives that may be trying to engage the same community in other participatory processes. This situation leads to a lot of strain in the communities and can also result in an overdose of participation. Identifying other ongoing initiatives, developing a communication link with them and looking for opportunities for synergy or collaboration should be part of the methodology.

These activities with a local community also allow researchers and practitioners to identify possible partners who could be involved in the research or development process. It could be a rural radio, a theatre group, or an NGO working with the same community. By establishing contacts from the outset, these groups will feel that they can play a useful role in designing the research initiative instead of perceiving themselves as mere service providers.

Local communities interact with governmental technical services, NGOs, development projects, rural media, community organizations and research organizations. All these organizations come to them with their own perspectives; in many cases, there is no link between the various development projects. To maximize the impact of the various local initiatives, it seems important to develop partnerships and build synergies at the community level.

This issue of collaboration is not an easy one. 'From Information to Communication in Burkina Faso: The Brave New World of Radio' by Ouattara and Ouattara raises the issue of collaboration with technicians from governmental services, and, more specifically, the problem of cohabitation of participatory and non-participatory approaches. Technicians are accustomed to executing and implementing programmes already identified by government authorities. Their mandate often consists in making people adopt their recommendations and participate in their programmes, which contradicts the participatory approaches that we want to implement. Therefore, there is a need to provide training in participatory development communication for the partners with whom we want to collaborate.

Constraints and challenges

If the foregoing steps are to be achieved, certain conditions must be met. In Part III, El Dabi gives an example from Egypt where participatory communication could not be introduced (see 'Introducing Participatory Development Communication within Existing Initiatives: A Case from Egypt'). This initiative aimed to develop implementation mechanisms for a strategic development plan in southern Egypt. The barriers that needed to be considered had to be identified, and realistic modifications that would enhance the potential participation of public, private and civil society actors in local development were to be proposed.

Local authorities were to be trained in participatory planning and participatory development communication, which was to be introduced by undertaking an assessment of the communication problems, channels and materials of all stakeholders; by designing a training programme for the immediate stakeholders to understand and apply the methodology in their communities; and by providing assistance to the immediate stakeholders to develop a strategy for their communities' development plans.

However, several obstacles hindered the implementation of this plan. First, participation was perceived as a process to allow stakeholders to voice their problems, not as a mechanism for them to look for ways of overcoming these problems. Second, the project did not allocate sufficient time to perform communication assessments or to conduct the training in a participatory way. Third, but not least, insufficient resources were allocated for the institutionalization of participatory approaches. As a result, it was not possible to introduce participatory communication in this particular context.

The chapter by Sow and Adjibade, also in Part III, provides examples of some of the practical difficulties we face when implementing participatory communication, particularly in a rural context (see 'Experimenting with Participatory Development Communication in West Africa'). This chapter also outlines some of the conditions that must be met. The authors raise the importance of prior knowledge of the local language and of the communication channels and tools used in the community; of negotiating with the men in a community to identify the conditions under which women can participate in specific activities; of time and distance considerations; of the development of partnerships with local organizations; of considering local authorities (traditional, administrative and family); and of harmonizing the understanding of participatory communication among the facilitators, decision-makers and participants involved. The chapter also reminds us that more time must be allocated to implement participatory communication processes than that usually planned in development projects.

Sow and Adjibade also remind us that participatory communication activities usually lead to the expression of the need for material and financial support to implement the solution(s) identified during the process. Provision must be made somewhere to answer these needs, whether as part of the initiative itself or through partnerships, otherwise the process stops where it should begin. The chapter shows that it is not useful to separate participatory communication activities from development activities and that resources must be planned to support these two complementary aspects.

Another chapter in Part III, 'Strategic Communication in Community-Based Fisheries and Forestry: A Case from Cambodia', presents the experience of introducing communication within a participatory natural resource management project in the Tonle Sap region of Cambodia. The initiative, as described by Thompson, emphasized communication as an integral part of its activities. It applied a wide range of tools and methodologies to inform, educate and promote participation. However, in the absence of a global communication plan, these efforts remained limited. Participatory development communication approaches can identify the best-suited community interventions and the management options for each community to ensure community-based natural resource management. However, the different communication activities must be integrated within a systemic and strategic plan to achieve their potential effectiveness.

Supporting natural resource management through communication strategies and tools

With PDC, communication strategies are developed around an initiative that has been identified by the community in order to tackle a specific problem or to achieve a common goal.

After community members have gone through the process of identifying a concrete initiative that they want to carry out, the next step is to identify both the various categories of people who are most affected by this NRM problem and the groups who might be able to contribute to the solution. They may be either specific community groups or other stakeholders who are, or could be, involved.

Addressing a general audience such as 'the community' or 'the farmers' does not really help to involve people in communication. Every group who makes up the community, in terms of age, gender, ethnic origin, language, occupation, and social and economic conditions, has its own characteristics, its own way of perceiving a problem and its solution, and its own way of taking actions. Likewise, communication needs will vary considerably within each specific community group or stakeholder category.

In all cases, it is important to pay particular attention to the question of gender. In every setting, the needs, social roles and responsibilities of men and women are different. The same is true of the degree of access to resources, participation in decision-making processes and the way in which they will perceive a common problem or potential solutions. The same is true for young people. There is often a sharp distinction between the roles and needs of girls and older women, as well as between the perceptions of older men and young people in the face of the same problem. Consequently, their interests are different, their needs are different, the ways in which they see things are different, and their contribution to the research or development initiative will also be different.

Communication needs and objectives

Development needs can be categorized broadly into material needs and communication needs. Any given development problem, as well as the attempt to resolve it, will present needs related to material resources and to the necessary conditions to acquire and manage these resources. However, there are complementary needs that involve communication in order to share information, influence policies, mediate conflicts, raise awareness, facilitate learning, and support decision-making and collaborative action. Clearly, these material and communication aspects should go hand in hand and be addressed in a systemic way by any research or development effort.

Nevertheless, participatory communication puts a greater focus on the second category of needs and ensures that they are addressed, together with the material needs that the research or development effort is concentrating on.

Communication objectives are based on the communication needs of each specific group concerned with the identified problem or set of activities that will be part of the project. These communication needs are put together by all the stakeholders involved and go through a selection process. The choices can be made on the basis of the needs that are most urgent or those that are most susceptible to action. These needs are then translated into a series of actions that should be accomplished to address each requirement.

Generally, in the context of NRM, these actions are linked to one or another of the following communication functions: raising awareness; sharing information; facilitating learning; supporting participation, decision-making and collaborative action; mediating conflicts; and influencing the policy environment.

Using communication tools in a participatory way

We often find situations where researchers or practitioners who want to use communication in their activities intend to produce a video, a radio programme or a play without first trying to identify how it will contribute to the research or development initiative. The expression 'communication tools' in itself implies that they are not the 'product' or the 'output' of the communication activities.

Participatory development communication takes another perspective. It leads participants through a planning process, which starts with the identification of the specific groups as well as their communication needs and objectives. The research or development team, together with community members and other stakeholders, then identifies the appropriate communication activities and tools that are needed to reach these objectives. It is a collective and consensus-building process, not a strategy developed outside the social dynamic.

PDC also puts traditional or modern media on the same level as interpersonal communication and learning experiences, like field visits or farmers' schools. The importance of using these communication tools in a way that will support multiple-way communication must, of course, be clearly stated at the outset of the project.

We have to consider two situations regarding communication tools. We often discuss this issue with the perspective of research and development teams using communication tools to support their activities. However, community members must also be able to use these communication tools for their own purpose.

Three criteria seem particularly useful in selecting communication tools: their current use in the community, the cost and constraints of their use, and the versatility of their uses. Whenever possible, we should first rely on the communication tools already in use in the local community for exchanging information and viewpoints, or the tools people are most comfortable with. Considerations of cost and sustainability and of different kinds of use should also be examined before making a decision.

In terms of the tools used by NRM practitioners, the chapters in this volume highlight various mixes of interpersonal communication and community media: community thematic discussions, participatory theatre, radio and participatory communication, farmers' field schools, video, photography, illustrations and community meetings.

Community thematic discussions

Almost everybody considers community thematic discussions as an important communication tool. But these discussions also imply a process and specific attitudes on the part of the facilitator.

Thiamobiga's chapter 'How the Parley Is Saving Villages in Burkina Faso' in Part IV gives us two examples of facilitators and the processes that are at work when using this tool. The chapter also describes a specific case in which community discussions were instrumental in managing bushfires and preserving the natural environment. Thiamobiga stresses the link between participatory communication and the '*parley*' – a traditional way of addressing issues and problems at the community level.

Participatory theatre

Participatory theatre also appears to be a favorite communication tool. In 'Burkina Faso: When Farm Wives Take to the Stage', also in Part IV, Thiamobiga discusses the experiences of a theatre of women farmers and explains the process of using theatre debate as a participatory communication tool. Theatre debate, in which a discussion follows the play and some parts of the theatre are played repeatedly following comments, was used to tackle soil fertility problems and was employed by women as a form of empowerment. The idea, at first, was to use this tool to help women voice their concerns and to illustrate causes and potential solutions associated with the problems. But the process initiated an empowerment process through which the women decided that they would play themselves.

There is a traditional ceremony performed in times of drought, when women are allowed to dress up as men to call for rain. The participating women wanted to refer to that ceremony so that they could bring forward topics that could be addressed directly by the men of the community without the risk that they would be offended (during the ceremony, men do not have the right to take offence).

By participating in the discussion to identify the problems related to soil fertility and by learning to express themselves as actors in a play, the women not only put their plots' soil fertility problem on the community agenda, but also gained self-confidence and became more assertive. The impact was also stronger because, in this case, community members were addressing other community members about common issues, rather than development actors from the outside promoting solutions.

At the same time, such involvement from community members, in this case women farmers, raised expectations that could not be met after the completion of the intervention. There was no direct follow-up, and although the experience was empowering for the participants, there was little impact at a broader level. This issue addresses the importance of planning for scaling up a specific intervention at the very beginning of the planning phase.

Radio and participatory communication

Ouattara and Ouattara's chapter 'From Information to Communication in Burkina Faso: The Brave New World of Radio' in Part IV reminds us not only that radio is the most common media in rural Africa, but also that it is still underdeveloped as a participatory communication tool. The research first started to use radio to promote the involvement of community members, together with a communication strategy based on 'endogenous communicators'. The programmes were designed on the basis of interviews and discussions conducted with community members and a communication team that included a radio producer, a representative of the farmers and a representative of a development structure active in the region (e.g. a development initiative or NGO). The development representatives were trained to prepare the field activities, to participate in the production of programmes and to collect feedback following the broadcast.

Other activities were then introduced to complement the media approach and to reinforce community participation. NRM problems and potential solutions were identified through discussion groups consisting of women, young people and adult men. In each locality, a committee which included local development actors was set up to define activities that could respond to prioritized needs. At the village level, a communication committee was involved to facilitate implementation. These field activities were then used in the production of radio programmes that were broadcast by the local rural radio station between two field trips. Essential questions asked by community members were discussed in these programmes. Specialists would also comment on these questions and participate in a dialogue with community members.

These activities have opened up a space for dialogue in the communities about natural resource management problems, while promoting synergy between the different development actors working in the same locality. The decisions resulting from this dialogue and the exchanges of information have involved community members and engaged them in a process in which they actively search for solutions instead of passively waiting for external assistance (e.g. by getting rid of pest-infesting orange trees, by resuming a dialogue between farmers and pastoralists and by enabling women to have a voice at community meetings).

Nevertheless, this experience also showed the difficulties associated with a participatory approach – namely, the danger of raising expectations without the possibility of responding to identified needs. For example, after prioritizing the lack of access to drinking water in the locality, community members and the team did not have many solutions to offer because the communication initiative was not associated with any specific development action or equipped with a structure that had the technical and financial resources to answer those needs.

In 'And Our "Perk" Was a Crocodile: Radio Ada and Participatory Natural Resource Management in Obane, Ghana', Larweh describes a situation in which a community was confronted with a decision to either migrate or renew its waterway, which was now choked by weeds, trees and debris. In fact, it no longer existed for most of the year. The community radio was part of a process where the community discussed the situation and decided to clear 40 years of accumulated debris. Neighbouring communities joined in the collective work. Four years later, the river irrigates the fields and is navigable. Through participatory communication, the community was able to unite around a single goal and transform their situation using their own means.

Video, photography, posters and brochures

In other situations, especially those aiming at empowerment, community members will take the lead in using communication tools or in taking decisions regarding the design, production and use of communication materials. Community input is well documented in Odoi's chapter on introducing villagers to video production, photography and the making of posters and brochures (see 'Communication Tools in the Hands of Ugandan Farmers' in Part IV).

The chapter describes the experience of farmers who edited a video that had been produced by the research team to share the results of their activities with other farmers. In this case, the farmers rejected the video because they were convinced that they could do a better job in delivering their own messages and experiences. The farmers first had a meeting to decide who should show what and how, set a date for the new recording, and signalled to the researchers when they were ready. This would have never happened if the researchers had not undertaken a process of participatory communication with the farmers – a clear manifestation of the farmers' empowerment.

The same thing happened with the photographs. After the pictures were developed, the farmers rejected them and started anew. For some time, research team members were discouraged and wondered when the production process would end.

As for printed materials, Odoi's chapter explains that the farmers could easily produce a brochure, but that the production of the poster was more difficult because this was a new concept for them. On examining a poster depicting proper water and sanitation practices placed at the entrance of the community hall, farmers said that it was teaching someone how to write. Clearly, the tool was not adapted to this specific community.

Tools should also be considered from the viewpoint of their usage. In a case from Lebanon by Hamadeh et al (see 'Goats, Cherry Trees and Videotapes: Participatory Development Communication for Natural Resource Management in Semi-Arid Lebanon' in Part III), a video and a local users' network inspired by a traditional way of communicating and resolving issues were used to manage conflicts and to facilitate the expression of views by marginalized people.

This research focused on understanding changes in resource management systems in an isolated highland village that was in the process of moving from a traditional cereal livestock-based economy to a rain-fed stone-fruit production system, and of improving prospects for sustainable community development. Community members were involved at different stages, and capacity-building was sought by establishing a local users' network.

The network acted as a medium to bring together the different users (e.g. cherry growers, flock owners and women), researchers, development projects, government officials and representatives of traditional decision-makers. It also supported participation, communication and capacity-building efforts.

The project used a traditional way of communicating and resolving dilemmas called *majlis*, in which issues are brought up within the community. As the network grew, so did the researchers' understanding of communication principles and the need to develop specialized sub-networks. Three sub-networks were developed, two of them dealing with the main production sectors in the village (livestock and fruit growing), while a third one addressed women's needs.

Tools and practices were mainly interpersonal: round-table meetings, community outreach by students, joint field implementation of good NRM practices, and workshops on different NRM themes. Short video documentaries were also produced and used during meetings.

Video was experimented with as part of an effort to involve the community in dialogue and conflict resolution. Marginal groups could express their points of view and the images helped to shed light on some aspects of conflict and dissent. The videos were shown in the presence of all parties, and the showings were followed by discussions that were also filmed and documented. A revised video that included the earlier discussions was then shown to the whole village until a positive dialogue started to emerge from the audience. Video was also used to highlight the economic productivity of women and to prompt discussions.

It was found that video helped marginal groups, who were usually shy in formal meetings, to express themselves. Videos were also found to be useful in generating discussions and awareness among and between different people and factions.

Influencing or implementing policy

Promoting poverty alleviation, food security and environmental sustainability also requires changes in the institutional and legislative environment. Local and national authorities, policy-makers and service providers all contribute to shaping and enforcing the regulatory environment in which the required changes must take place. Therefore, it is important to facilitate dialogue at that level in order to gain support for the initiatives developed by local communities.

Facilitating dialogue at the level of local and national authorities, policy-makers and service providers, or between communities and the policy environment, or advocating changes in the policy environment, are other roles for the NRM practitioner or researcher as a communication actor.

Two chapters from Cambodia in Part III ('Communication Across Cultures and Languages in Cambodia' by Kimhy and Pinreak and 'Talking with Decision-Makers in North-Eastern Cambodia: Participatory Development Communication as an Evaluation Tool' by Kimhy) provide examples of how participatory communication can influence policy and help with its implementation.

In 'Talking with Decision-Makers in North-Eastern Cambodia: Participatory Development Communication as an Evaluation Tool', Kimhy shares the experiences of indigenous communities who evaluated a natural resource management initiative implemented by the government and presented their findings to government officials. The presentation also included recommendations to the government in a context where government representatives usually tell communities what they should do. In this case, evaluation was used both as an empowerment tool for community members and as an advocacy tool for influencing the government.

In 'Communication Across Cultures and Languages in Cambodia', Kimhy and Pinreak describe a situation in which a team was visiting villagers, in the context of a new land rights extension law, in order to inform them of their rights and of existing laws. Transferring information or knowledge across cultural and language barriers is difficult; but it is much more difficult when some of the concepts do not even exist in the vocabulary of the people with whom we want to converse. This was the case in this initiative because concepts such as laws and land titles did not exist for the indigenous communities with whom the team was interacting. At the same time, communicating these concepts was an important task because powerful interests were taking away their lands and forests, and the communities did not know what to do.

At first, the team, who did not speak the indigenous languages and had prepared its information material without any community involvement, failed to reach the communities. It then experimented with a participatory communication approach, involving community members in preparing the sessions and communication material. It also included indigenous people as full members in the land rights extension work, which changed the team's whole approach to working with communities.

It is interesting to note that the team also used a tool called the 'Livelihood Framework' in the course of its discussions with the communities. It presented ideas expressed by the community in pictures that were painted and then revised by the community. The visuals, in this case, greatly assisted in the discussions and expressions of different viewpoints.

A chapter from the Philippines by Torres (see 'Paving the Way for Creating Space in Local Forest Management in the Philippines' in Part III) tells how participatory communication helped to implement community-based natural resource management with indigenous people. When community-based forest management was adopted as a national strategy in the Philippines, issues started to emerge with regard to the readiness and capacity of communities to handle the delegated tasks and functions.

On its part, an upland people's organization called the Bayagong Association for Community Development was able to assert, legitimize and sustain its control over a piece of forestland that it had been occupying *de facto* for years. To do so, community members underwent a process of participatory resource management planning. In the gender-sensitive PRA methodologies that were used during a year of action research, participatory communication was a core process.

This participatory experience helped participants to get a better grasp of their resource quality, to assess their own capacities and weaknesses, and to identify internal and external threats, as well as how these could be handled. It enabled them to gain the knowledge, attitudes and skills that were necessary to deal with the management of their forestland and to develop rational approaches to forest management. But they also learned to become more open and assertive about their rights, and soon became empowered in identifying and addressing their needs by using locally available resources before turning to outside sources for assistance.

Participatory development communication played a critical role in tempering the socio-political environment so that a climate favourable to the community's takeover of the forestland be created. However, success was not only due to communication. Other factors such as social capital, policy presence and assistance by external actors also played a role. What is unique is that participatory communication enabled the evolution of a 'participation-as-engagement' process veering away from the usual 'participation-as-involvement' process.

A chapter from Indonesia by Jahi (see 'From Resource-Poor Users to Natural Resource Managers: A Case from West Java' in Part III) tells of a research project that originated from a question that researchers asked themselves while they were doing a baseline study in a remote rural area. The researchers wondered whether poor farmers and landless farm labourers could participate in managing a strip of public land that stretched out along a river and thus be able to derive benefits from that activity.

By law, farming activities were prohibited on this land. Only grass cultivation was allowed on the riverbanks, which had been raised to prevent local flooding. However, regardless of the regulations, landless farmers continued their farming activities on the riverbanks. Meanwhile, officials of the Department of Public Works kept enforcing the regulations and eradicating the crops. A consensus then developed. The farmers could continue their activities provided that they grew grass at least 1m away from the river's edge. Sheep rearing was encouraged.

The researchers established links between university researchers, local government officials, extension services, village governments and local farm communities. Communication materials such as slide shows, posters and a comic book were developed and tested with farmers and extension workers. Different topics were developed for different audiences. For example, presentations on the potential of raising sheep were prepared for local policy-makers and decision-makers, and aspects of sheep production and rural family budgets were covered in productions for extension workers and farmers.

Capacity-building for livestock extension workers and farmer leaders was then provided. In-kind loans in the form of sheep were provided to the farmers, who agreed to return a certain number of the offspring to the project. Supervision and backstopping activities were also provided to farmer leaders, who agreed to share the information with other farmers after they had acquired enough experience.

Farmer-to-farmer communication was encouraged and supported. Indeed, it was found to be a more efficient way of raising farmers' interest than the methods that researchers or extension workers used to employ. The experience also raised public and private interest in supporting economic activities, such as sheep rearing, in the district. Fifteen years after the beginning of the project, livestock production in the district has developed significantly and small farmers can still earn their living in this way.

In the policy arena, there are also situations where participatory communication must coexist with inadequate policies and help to seek solutions. In a chapter presenting the case of the Kahuzi-Biega National Park in the Democratic Republic of Congo (see 'Conserving Biodiversity in the Democratic Republic of the Congo: The Challenge of Participation' in Part III), we find a situation in which a conservation measure (the creation of a park to protect a unique ecosystem and a population of mountain gorillas) was implemented in a top-down way without the involvement of the population living in, or on the fringes of, this newly protected territory. In such a conservation model, the population is excluded from managing natural resources. Consequently, local people do not participate in or support the new unpopular measure.

In this case, an alternative plan had to be developed. Using environmental communication and in collaboration with the population living in the area, community development activities compatible with the conservation of the park and its natural resources began to be planned and implemented. These activities quickly led to the development of mechanisms for participatory management. Soon, some 200 village parliaments were set up to facilitate the process. Not only have local opinions changed towards the park, but the communities have also started taking charge of its protection.

The promotion of policies goes hand in hand with collective action. One chapter in Part III (see 'Giving West African Women a Voice in Natural Resource Management and Policies' by Ouoba) depicts the daily life of a rural woman of the African Sahel and her difficulties with regard to natural resources: lack of access to water and fuelwood; problems of soil fertility; and lack of access to landownership. It also tells of the efforts of a rural women's association to find collective answers to these individual problems. The solutions to the NRM problems experienced by rural women must come from their own efforts, a process that can be facilitated by participatory communication. We can see that such initiatives are part of an empowerment process in which marginalized people, who are not used to expressing themselves on such issues, develop confidence and learn to voice their difficulties and needs, and to formulate specific actions to address these needs.

Capacity-building issues

Participatory development communication and, more broadly, the use of communication in the context of participatory development or participatory research have to be appropriated by NRM researchers and practitioners. This should also be the subject of exchanges and discussions with other stakeholders who participate in these activities, such as community members.

Five chapters in Part V ('Forging Links between Research and Development in the Sahel: The Missing Link'; 'Isang Bagsak South-East Asia: Towards Institutionalizing a Capacity-Building and Networking Programme in Participatory Development Communication for Natural Resource Management'; 'Implementing Isang Bagsak in East and Southern Africa'; 'Implementing Isang Bagsak: Community-Based Coastal Resource Management in Viet Nam'; and 'Implementing Isang Bagsak: A Window to the World for the Custodians of the Philippine Forest') discuss the implementation of Isang Bagsak, a learning and research programme in participatory development communication. The expression 'Isang Bagsak' comes from the

Philippines and means reaching a consensus, an agreement. Because it refers to communication as a participatory process, it has become the working title for this initiative.

The programme seeks to increase the capacity of development practitioners, researchers and associated stakeholders who are active in the field of environment and natural resource management in using participatory development communication to work more effectively with local communities and associated stakeholders. It aims at improving the capacity of practitioners and researchers to communicate with local communities and other stakeholders, and to enable them to plan, together with community members, communication strategies that support community development initiatives.

The programme combines face-to-face activities with a distance-learning strategy and web-based technology. With the distance component, the programme can answer the needs of researchers and practitioners who could not easily leave work for a campus-based programme. It is currently implemented in South-East Asia and Eastern and Southern Africa, and is in the start-up phase for the African Sahel.

In South-East Asia, Isang Bagsak is being implemented by the College of Development Communication of the University of the Philippines at Los Baños. It works in the Philippines, Cambodia and Viet Nam.

In the Philippines, the programme is implemented in partnership with the Legal Assistance Centre for Indigenous Filipinos (PANLIPI), an NGO devoted to legal assistance to indigenous Filipinos. The goal of the programme is to bring indigenous people into the mainstream of learning about NRM.

In Viet Nam, capacity-building in participatory development communication aims to improve approaches to coastal resource management, understand how to influence local policies in order to improve participatory management of coastal resources, and form a national network in community-based coastal resources management. Furthermore, a Vietnamese version of the programme, called *Vong Tay Lon*, is being prepared.

In Cambodia, participants come from the new forest administration department. This national body is responsible for formulating and implementing forest policies, which affect more than half the country's total land area. Its mandate includes the elaboration of a statement on National Forest Policy, which will be based on a consultative process inclusive of all stakeholders in national forestry policy formulation.

In Southern and Eastern Africa, the programme is implemented in Zimbabwe, Malawi and Uganda by the Southern Africa Development Community Centre of Communication for Development (SADC-CCD). By building capacity in participatory development communication, the programme aims to facilitate collaboration among decision-makers, planners, development agents and communities in order to improve the management of the environment and natural resources, as well as the research and development initiatives. The programme works in partnership with the National Agriculture Research Organization in Uganda, with a national resource management research initiative called the Desert Margins Initiative in Malawi, and with the Department of Agricultural Research and Extension in Zimbabwe.

Another programme is being initiated with an agroforestry network in Senegal, Burkina Faso and Mali, which will be led by the Sahel Programme of the International Centre for Research in Agroforestry (ICRAF-Sahel). In the Sahel, the starting point for implementing Isang Bagsak is the realization that new agroforestry technologies which should improve lives are not being widely adopted despite efforts to convince people to do so. The objective of the programme is to reinforce the capacities of the different actors so that they can co-produce and co-disseminate new knowledge in collaboration with all involved stakeholders.

El Hadidy addresses the issue of capacity-building in the context of the Arab region, but situates participatory development communication within the larger framework of participatory development (see 'Reflections on Participatory Development and Related Capacity-Building Needs in Egypt and the Arab Region', also in Part V). This chapter advocates that practitioners should engage in a critical reflection of their practices. It states that, in itself, the 'delivery of resources' mode of operation in the form of transfer of know-how and skills is not sufficient. It also indirectly implies that resources are transferred from those who have them to those who do not, instead of recognizing that every practitioner has skills and abilities that need to be brought to the surface. Unlike capacity-building that requires a 'how-to' approach, such as proposal writing or business planning, capacity-building in participatory development communication should focus on recognizing that communication is a natural process. It advocates an approach based on the facilitation of resourcefulness rather than providing resources. This process goes hand in hand with the documentation and discussion of local participatory practices.

The chapter by Thompson and Acunzo from the United Nations Food and Agriculture Organization (FAO) (see 'Building Communication Capacity for Natural Resource Management in Cambodia', Part V), presents a national capacity-building effort in Cambodia that was designed to help a communication team (which brought together the staff of two ministry communication units) design and implement targeted information and communication interventions to support plans and efforts made by local communities for natural resource management. The strategy was based on implementing information and communication strategies at the field level and providing in-service training in pilot sites. The learning process included participatory analysis, training of villagers, material design and production, as well as monitoring and evaluation for improving agricultural and fishing practices. The chapter describes the constraints and lessons learned in the course of this initiative. Among the challenges encountered, the authors mention that the lack of operational budgets makes it difficult for the newly trained communication team to apply their new skills. Similar trends have also been observed in other capacity-building initiatives. We need to address this situation as part of capacity-building efforts and to examine how the latter can be better integrated within the operational plans of targeted institutions.

Finally, capacity-building and co-learning efforts should also document and promote a systematic use of participatory development communication in natural resource management.

It is important to state that there is no single all-purpose recipe to start a participatory development communication process. Each time, we must look for the best way to establish the communication process among different community groups and stakeholders, and use it to facilitate and support participation in a concrete initiative or experimentation driven by a community to promote change.

However, participation in the planning process is important. We already stressed that using participatory development communication demands a change of attitude from researchers and development practitioners. Traditionally, the way in which many research teams and practitioners used to work was to identify a problem in a community and experiment with solutions, with the collaboration of local people. On the communication side, the trend was to inform and create awareness of the many dimensions of that problem and of the solution that community members should implement (from an expert viewpoint). This kind of practice has had little impact; but many researchers and development practitioners still work along those lines.

Working with participatory development communication means involving the local community in identifying the development problem (or a common goal), discovering its many dimensions, identifying potential solutions (or a set of actions) and making a decision on a concrete set of actions to experiment with or implement. It is no longer the sole responsibility of the researcher or development practitioner.

Participatory development communication supports a participatory development or research-for-development process. Such a process is usually represented through four main phases, which, of course, are not separated but flow into one another. These stages are problem identification, planning, implementation, and monitoring and evaluation. At the end of the process, a decision is made that either consists of going back to the beginning of the process (problem identification) and starting another cycle or revising the planning phase, or scaling up efforts and starting another planning, implementation and evaluation cycle.

The participatory development communication model supports this process. Bessette (2004) discusses the most common steps when planning and implementing participatory development communication in a natural resource management context:

- Step 1: establish a relationship with a local community and understand the local setting.
- Step 2: involve the community in identifying a problem and potential solutions, and in carrying out a concrete initiative.
- Step 3: identify the different community groups and other stakeholders concerned with the identified problem (or goal) and initiative.
- Step 4: identify communication needs, objectives and activities.
- Step 5: identify appropriate communication tools.
- Step 6: prepare and pre-test communication content and materials.
- Step 7: facilitate the building of partnerships.
- Step 8: produce an implementation plan.

- Step 9: monitor and evaluate the communication strategy, and document the development or research process.
- Step 10: plan the sharing and utilization of results.

The process, nevertheless, is not sequential. Some of these steps can be implemented in parallel or in a different order. They can also be defined differently depending upon the context. It is a continuous process, not a linear one. But the steps can guide the natural resource manager researcher or practitioner in supporting participatory development or research through the use of communication.

Institutional aspects

Implementing participatory development communication faces the same constraints as the participatory development process that it supports: it demands time, resources and practical modalities that can only result from negotiating with the donor organizations involved.

Initiating the process

In traditional development culture, financial support often comes after the revision and acceptance of a formal proposal, whether it is a research for development proposal or a development project proposal. In order to go through the different levels of revision and acceptance, the proposal must be clear and complete. The development problem or goal must be clearly identified and justified, the objectives must be outlined with precision and all the activities must be detailed. The full budget, of course, must figure in the proposal with all its budget notes.

Although some organizations are rethinking the process and promoting programme instead of project orientation, in most cases this is the situation we face. It is important to put this issue on the agenda of donor organizations and to demand a review of the procedure: if we want to develop a participatory development process and have community members and other stakeholders have their say during all phases of the process, starting with the identification and planning phases, this means that we need the time and resources to do so.

In the meantime, we can identify two modalities that can be proposed to the donor organization. The first consists of putting together a pre-proposal that will seek to identify and plan the project with all stakeholders. The second modality – which is really a second choice in case the first one is not possible – consists of building the proposal in a way that will permit its revision with community members and other stakeholders.

Changes during implementation

A participatory development or research process cannot be planned in the same way as the construction of a road: as participation is facilitated and more feedback is gathered, a wider consensus develops and decisions are made; things change. This is why it is always an iterative process and we must have the possibility of changing plans as we go along in order to attain the objectives that have been identified.

This must also be discussed with the donor organizations involved since traditionally, once a proposal has been accepted, nothing can be changed.

Time considerations

The length of activities is also a problem. Proposals often have to be developed within a two- or three-year time frame. But participation takes time, and in many cases, this span is barely enough to really start the process. Thus, even if the expected results have not yet happened, it is necessary to identify the progress made by the research and development activity and to build the case for the continuation of support. This also underlines the importance of a continuous evaluation mechanism when implementing the process.

Regional perspectives

Two chapters in Part II, from Africa and Asia, examine participatory development communication from a regional perspective.

In Asia, Quebral, who was the first to use the term 'development communication' more than 30 years ago, retraces the evolution of participatory approaches to development communication (see 'Participatory Development Communication: An Asian Perspective'). The chapter situates this evolution in the context of the communication units, departments and colleges in Asian universities and from the perspective of fighting poverty and hunger. Quebral notes that development communication does not identify itself with technology *per se*, but with people, particularly the disadvantaged in rural areas. Participatory development communication uses the tools and methods of communication to provide people with the information they need and to reinforce their capacity to make their own decisions.

The chapter insists on recognizing the beginnings of development communication and on expanding upon earlier achievements. Older models retain their validity in certain situations and can still be used when appropriate. The chapter also presents lessons and observations learned through this Asian experience.

In the context of natural resource management, Quebral insists on the importance of a balance between technology and people's empowerment, and on how participatory development communication can help people to zero in on their problems. It can also support people's choice of the technologies that they wish to experiment with.

Offering another regional perspective, in 'Participatory Development Communication: An African Perspective' Boafo describes and analyses the application of participatory development communication within the African context and stresses the linkages between communication and the different dimensions of development on the continent.

Since the 1960s and 1970s, many development communication strategies and approaches have been used in numerous development programmes by a large number of development organizations. However, greater efforts remain to be made to address the constraints facing the practice of participatory development communication, particularly in the context of rural and marginalized communities, where the majority of the populations in most African countries reside.

In this context, notes Boafo, community communication access points and traditional media are of particular importance. Effective applications of participatory development communication approaches and strategies at the grassroots and community level should necessarily involve the use and harnessing of these communication resources. With their horizontal and participatory approaches, they can contribute effectively to enhancing participation in cultural, social and political change, as well as in agricultural, economic, health and community development programmes.

Conclusions

In the field of natural resource management, participatory development communication is a tool that reinforces the processes of participatory research and participatory development. It aims to facilitate the participation of communities in their own development and to encourage the sharing of knowledge needed during these processes. It integrates communication, research and action within an integrated framework. Furthermore, it involves researchers, practitioners, community members and other stakeholders in the different phases of the development process. But most importantly, participatory development communication points out that natural resource management must be directly linked to the agenda of communities and must seek to reinforce their efforts to fight poverty and to improve their living conditions.

For communication to be effective in addressing the three interlinked development challenges of poverty alleviation, food security and environmental sustainability, it must fulfil the following functions: ensure true appropriation and ownership (not just the buy-in) by local communities of any NRM research or development initiative; support the learning needed to realize the initiative and facilitate the circulation of relevant knowledge; facilitate the building of partnerships, linkages and synergies, with the different development actors working with the same communities; and influence policy and decision-making processes at all levels (family, community, local and national).

To achieve these objectives, a major effort is required in capacity-building – more specifically, in participatory learning – for practitioners in the field of natural resource management. Development workers, NGOs, researchers, extension workers and government agents responsible for technical services need appropriate communication skills. The ability to work with local communities in a gender-sensitive and participatory way, to support learning processes, to develop partnerships with other development stakeholders and to affect the policy environment should be recognized as equally important as the knowledge needed to address technical issues in natural resource management.

At the same time, field practitioners, researchers and community members who are involved in natural resource management initiatives are experienced in the use of communication within participatory research and development initiatives. There is no recipe that can be used in all situations; but there is much to learn from sharing, discussing and reflecting on our own experiences. As advocated in El Hadidy's chapter 'Reflections on Participatory Development and Related Capacity-Building Needs in Egypt and the Arab Region' in Part V, we should use an approach that facilitates resourcefulness rather than provides resources.

Of course, such a process goes hand in hand with the documentation and discussion of our natural resource management and participatory development communication practices. This is why initiatives such as the Isang Bagsak programme and the FAO initiative in Cambodia should be developed, supported and multiplied in various contexts and situations. It is also why participatory learning in participatory development communication for both practitioners and stakeholders should be on the agenda of every organization supporting NRM research and development initiatives. It is only through such efforts that we can make participatory development happen, not only at the level of our discourses, but where natural resource management occurs in the field. It is also through such efforts that we can make sure that local actions can have a global impact by influencing the policy environment and making the knowledge available to those who really need it.

Finally, it is through such efforts that we can promote and cultivate the values which are at the core of our work, including the principle that people should be able to participate fully in their own development. In a recent report, Nora Quebral (2002) insisted that:

> *We now need to explicate those values more finely and cultivate them more rigorously in our actions. Our training procedures may have overly stressed skills at the expense of values. We need to make values more explicit, to deliberately pair them with the corresponding skills, if necessary. My first challenge, then, to development communicators is to make development communication values more pronounced in their practice.*

The same challenge can be extended to NRM practitioners and researchers: we need to make participatory development happen if we are to support communities and governments in their efforts to address the three interlinked development challenges of poverty alleviation, food security and environmental sustainability. Participatory development values, local and modern knowledge in natural resource management, and communication skills must be combined for this to happen.

References

Bessette, G. (2004) *Involving the Community: A Facilitator's Guide to Participatory Development Communication*, IDRC, Ottawa, Canada, and Southbound, Penang

Quebral, N. C. (2002) *Reflections on Development Communication (25 Years Later)*, College of Development Communication, University of the Philippines at Los Baños (UPLB), Los Baños, the Philippines

II

Regional Perspectives

Participatory Development Communication: An Asian Perspective

Nora Cruz Quebral

Asia is a region of many faces. This chapter speaks of the Asian experience with participatory development communication (PDC) from the perspective of one of its sub-regions – that grouping of nations known collectively as South-East Asia. More precisely, the chapter delimits itself to PDC as interpreted by communication units set up in South-East Asian colleges and universities as part of their agricultural extension or outreach function. The affinity of these units with the media offices of extension services in US land grant colleges[1] is obvious. Nonetheless, they have evolved – and continue to do so – into hybrid structures more appropriate to their cultures and to the state of knowledge in the field of development and communication.

There are other Asian viewpoints on PDC, notably in India and other parts of South Asia. They will be similar to the South-East Asian experience in some ways, different in others. All have lessons to offer in the continuing delineation of the relationship between communication and human development.

University communication units

The communication units referred to are found in Indonesia, Malaysia, the Philippines, Thailand and Viet Nam. They are at the incipient stage in transitional societies such as Cambodia, Laos and, perhaps, Myanmar. Regarded as adjuncts to the biological and physical science departments of their universities, the older units were initially tasked with extending the results of research generated by those departments, with some public relations and publicity jobs for administrators thrown in. This they were expected to accomplish through the media; hence, they were staffed with writers, editors, artists, audio and video specialists. Face-to-face interaction with farm families was considered something that extension workers do and, therefore, was outside the mandate of the communication staff.

An obsolete, ante-millennium model of PDC, you say? It is alive and kicking in South-East Asia in spite of globalization, state-of-the-art information and communication technology, participatory communication activism, terrorism and all other change-inducing phenomena now sweeping the world. Evidence of its endurance may well be mirrored, in greater or lesser degree, among the organizations usually represented in international events: the focus on proffered technology, the sidelining of communication practitioners within the organization, the forced merger of communication with other seemingly related units for reasons of efficiency, economy or whatever.

Evolution of participatory development communication in South-East Asia

There is another side to the picture, however. It was in this type of communication unit that PDC as study and practice first saw light in South-East Asia, was nurtured and then diffused to other developmental fields such as health and the environment, among others. At least seven of those university communication units have evolved into fully fledged teaching departments with their own research and outreach programmes. One has even achieved college status, although still under rather shaky circumstances at the moment.

Every forward step has meant greater latitude to break away from traditional characterizations and to chart their preferred direction while expanding their influence. In the College of Development Communication (CDC) at the University of the Philippines at Los Baños, for instance, the staff remains concerned with the agricultural content of PDC, but in the broader context of natural resource management (NRM) through their association with the United Nations Food and Agriculture Organization (FAO) and the

International Development Research Centre (IDRC), or of reproductive health through their projects with the Philippine Department of Health and the Johns Hopkins University Center for Communication Programs. Through formal and non-formal training programmes, CDC has produced hundreds of development communicators who have fanned out to other fields besides agriculture and to other countries outside South-East Asia. Through its various curricula and publications, then, reinforced by its links to research and action programmes such as Isang Bagsak,[2] CDC has become the nucleus of a major network engaged in the study and application of communication principles in or for development.

The participatory character of development communication has always been considered a given in most of South-East Asia, although the type and degree of participation may not always have been uniform. Until recently – for example, in Malaysia – 'participatory' did not always translate into direct critiques of government policies as in, say, the Philippines, where the political institutions are more Westernized – some would say too Westernized. On the other hand, even in an old democracy such as Thailand, participatory development communication as taught in the universities may still follow the top-down diffusion mode simply because of less exposure to ever-changing development communication thought as new insights are uncovered. As for a hierarchical society such as Cambodia's, particularly with its present form of government, participatory development is still uneven. There is less of it in formal communication encounters, but apparently a great deal more among peers in informal field settings. Clearly, PDC is a product of a society's culture, socio-political institutions and acceptance of current thinking in development and communication. It is also clear that PDC professionals everywhere have a great opportunity to enlarge the degree of citizen participation in their societies by always making visible in their work the principles of participatory development.

And what is the essence of PDC in South-East Asia today? Mindful of its beginnings, PDC aligns itself with those who would reduce, and possibly eliminate, hunger, poverty and sickness in the world. Yet, as a social science, it does not identify with technology *per se* but with the people who use or do not use it, particularly among the disadvantaged in rural areas. Thus, its ultimate goals are equality and social justice for all and freedom for everyone to develop their potential. It uses the tools and methods of communication chiefly to educate through non-formal ways so that people may have both the capacity and the information to make their own decisions.

Some observations and reflections

South-East Asian development communicators have had their problems and setbacks, to be sure. But they have also had their high moments and successes. Through it all they have learned from their own experience and that of others. A few of their more current observations and reflections on participatory development communication are shared below:

- The conceptual difference between communication as process and communication as media or channels seems to bear repeating every so often. As process, it is the exchange or interchange of all types and kinds of information within a society or social group, which is why communication is said to be the most basic of all social processes. It is also seen by many as communication through mass or community media. This traditional perception can be enlarged to include all the mechanical and personal avenues through which information flows between and among the members of a social group. Whether seen as process or as channel, communication can be consciously used for development. Development communication, then, is the process of multilevel exchange within a society of information whose intent is to advance human development and which is channelled through selected media.
- Communication media have long been dichotomized into mechanical and personal. Mechanical media, such as radio, television and now information and communication technologies (ICTs), have received much more attention. It is time to ferret out the nuances of interpersonal communication that promote development. Initiatives such as Isang Bagsak have made a good start. They can also explore the workable combinations of face-to-face and mediated communication that delineate process. In this way, they can take the concept of development communication process out of the generic stage that it is still in and give it more precision and specificity.
- It seems to have become standard in many disciplines for younger professionals to denigrate the work of their predecessors around the globe as being reactionary, perhaps forgetting that they do so from the vantage point of hindsight. And so they try to reinvent the wheel. Without the foundation laid down by those seminal thinkers around the world who have gone before, today's communicators, for instance, would not have had the concept of communication in, or for, development to begin with, and to which is now attached the 'participatory' label. A lesson worth sharing with other development professionals is this: do not turn your back on your beginnings. Acknowledge them, even as you build on them.
- In the practical realm, new models of communication do not necessarily replace older ones. They just co-exist. This is reflected today in the undiscriminating use of terms associated with both old and new models. As a case in point, 'target audiences' and 'beneficiaries' are spoken of alongside of 'stakeholders' and 'participants'. PDC professionals should set the example of being clear about the kind of communication they advocate and of adapting their terminology accordingly. At the same time, they should recognize that older models retain their validity in certain situations and can still be used where appropriate.
- It is now accepted that rural people and other disadvantaged groups have the right to participate in decisions affecting their lives. They need to be empowered – as the stock phrase goes – to realize their self-worth, and to have their opinions heard and factored into the development dialogue. The same can be said of another group in the development world

– the extension technicians, media practitioners and other rank-and-file fieldworkers. In the diffusion model of technology transfer, they are the faceless middlemen who connect the scientists to the local communities. In later communication models, they are hardly visible and are, perhaps, just as neglected. They need to be recognized, too, as valued participants in the development process and accorded equal rank with the other actors.

- As shorthand for innovation, technical content, improved practice or – in our case – natural resource management, technology is not a bad word. PDC professionals should make their peace with it. Development needs a balance between technology and people empowerment. Neither one by itself can go it alone. PDC can be a tool to help people zero in on their problems and apply the technology they wish, given an adequate array to choose from and the capability to make the choice.

- Still on the subject of balance, the trend seems to be for development communication as art and language to break new ground in areas where development communication as social science has only ventured peripherally. This is a welcome move for which some caveats may be offered. Unilateral answers have never worked before and there is no reason to believe that they will now. Development is a multifaceted pursuit and PDC practitioners must integrate within it as many facets as are feasible. On another note, anecdotal case studies without back-up systematic investigation could lead us back to equating PDC with its channels, whether mechanical or personal, rather than with it as process.

- Communicators have been accused of talking only to themselves. Should PDC professionals not also discuss overlapping concerns with researchers, practitioners and administrators involved in natural resource management? Many development professionals still operate within the old researcher–extension worker–farmer paradigm of technology transfer, perhaps because they have not been exposed to newer ones. PDC could facilitate that type of dialogue not only through mediated communication, but also at meetings in venues that NRM researchers, practitioners and administrators are familiar with.

- With the present state of world finances, many countries, including developed ones, can no longer support one-on-one intensive, but expensive, extension systems, potentially leaving the field to commercial companies. What alternatives does the richness of information and communication technology have to offer to small farmers with dissimilar needs? Isang Bagsak has piloted a possible community-based resource. More experiments like this are essential when done systematically and with an eye to their fiscal viability for poor countries.

- Is participatory development communication a means or an end, or both? Is Isang Bagsak meant to achieve better NRM in a community, or is it a way for researchers and community residents to internalize PDC? Or are both objectives valid? The answer will dictate what indicators should be used to gauge the success of projects such as Isang Bagsak.

- Finally, PDC can be institutionalized in two ways: in policy to ensure its adoption by field practitioners, and in theory to ensure its continuing viability and validity through research conducted by students and academics. Both will enrich communication for development as practice and as a field of study.

Conclusions

Participatory development communication, in its several variations across countries, is a young but dynamic field that is nurtured by many disciplines. At the same time, its unique window to human development allows it to pioneer new concepts and practices that other fields can emulate. It has come quite a way in the span of 30-odd years. Like science and art, it can contribute much more as long as its advocates, with their own kind of tools and expertise, hold fast to their vision of equality and social justice for all, and freedom for everyone to develop their potential.

Notes

1   In the US, land grant colleges are a set of state and territorial institutions of higher learning that receive federal support for integrated programmes of agriculture teaching, research and extension for agriculture, food and environmental systems.
2   Isang Bagsak is a learning and networking programme that aims to improve communication and participation among natural resource researchers, practitioners, communities and other stakeholders, and to provide communication support to development initiatives in helping communities overcome poverty.

Participatory Development Communication: An African Perspective

S. T. Kwame Boafo[1]

Since Nora Quebral (1971) of the University of the Philippines at Los Baños first used the term 'development communication', the concept has become entrenched in communication studies and practice in Africa. Several African communication scholars, researchers and practitioners have written about and attempted to contextualize the application of the concept in the African environment; a number of study materials have also been produced and virtually

all communication teaching and training institutions in the continent offer courses and programmes on the concept, often under such appellations as 'communication and development'; 'participatory development communication'; 'communication for development'; 'communication for social change'; and the older but more practice-oriented term 'development-support communication'.[2]

The concept and its different appellations have been very comprehensively defined in the communication literature and we will not attempt to do any in-depth definitional or operational analysis here.[3] A working definition that contains many of the tenets and assumptions of the concept, and which we will adopt in this chapter, refers to development communication as the planned and systematic application of communication resources, channels, approaches and strategies to support the goals of socio-economic, political and cultural development. Participatory development communication (PDC) puts accent on the process of planning and using communication resources, channels, approaches and strategies in programmes designed to bring about some progress, change or development, and on the involvement of the people or community in change efforts. As Ascroft and Masilela (1994) have aptly noted, in the African context just as elsewhere, participation translates into individuals being active in development programmes and processes; they contribute ideas, take initiative and articulate their needs and their problems, while asserting their autonomy.

This brief chapter describes and analyses the application of participatory development communication within the context of African countries. It attempts to situate the use of development communication within the social, economic, political and cultural development challenges and realities in the region and draws attention to a number of contextual factors that determine the effectiveness of participatory development communication in Africa.

Development challenges and development communication

The challenges in social, political, cultural and economic development and transformation in African countries are very well articulated in various documents, publications, conferences, plans of action and programmes, and are particularly well summarized in the recently launched New Partnership for Africa's Development (NEPAD).[4] Although it is not important here to enumerate the varied development challenges facing African countries, it is nevertheless relevant in the context of participatory development communication to stress that there are linkages between communication and the different dimensions of development in Africa, whether they are political, social, economic or cultural. Research studies and experience in diverse contexts and countries in Africa have clearly demonstrated that development communication approaches can be used to enhance participation in cultural, social and political change, as well as in agricultural, economic, health and community development programmes. In a word, regardless of the type of development challenges in African countries, there is some function for communication and information in the efforts made to address those development challenges.

Since the decades of the 1960s and 1970s, development communication strategies and approaches have been employed in numerous development programmes and projects across the length and breadth of Africa. A variety of development communication approaches and strategies have been used by international organizations, funding agencies, government departments, non-governmental organizations (NGOs) and civil society groups in development-oriented programmes and projects designed, *inter alia,* to improve agricultural production; tackle environment problems; prevent and manage health problems and pandemics such as malaria and HIV/AIDS; improve community welfare, the status of women and educational levels; promote or enhance democracy and good governance; and encourage local and endogenous cultural expressions and productions. But the practice of development communication in Africa has been done in the face of several major communication constraints in the region. These constraints are well documented in a number of publications and reports on communication in Africa[5] and include the following:

- weak and inadequate infrastructure and spread of communication and information systems, as well as limited financial resources to develop or strengthen them;
- dislocation and disparities in communication and information flows between urbanized areas and rural communities, as well as disadvantaged population groups, because of insufficient access for large segments of the national populations to modern communication and information means; and
- low priority given by policy- and decision-makers to communication and information as integral components of development programmes; this low priority often translates into the absence of effective policies and structures to guide, manage, coordinate and harmonize communication for development activities in virtually all African countries.

Considerable efforts are being made in African countries to address the above communication constraints and difficulties with financial and technical support from a number of United Nations agencies, international and regional organizations, multilateral and bilateral funding agencies, and professional bodies. These efforts have gone a long way towards enhancing communication and information infrastructure; strengthening communication capacities; nourishing the emerging independent and pluralistic media; increasing access to communication and information systems; and developing human resources in communication and information in Africa. The efforts have resulted in the rapid development of community radio stations in such countries as Cameroon, Ghana, Mali, Malawi, Mozambique, South Africa and Zambia; the development of community multimedia centres and tele-centres in Ethiopia, Mali, Mozambique, Senegal, Tanzania and Uganda; the growth of independent and pluralistic media structures in such countries as Botswana, Burkina Faso, Ghana, Kenya, South Africa and Tanzania; improved communication training programmes in a number of African countries leading to more professionally trained communication practitioners; and, along with the profound changes from monolithism to pluralism in the political landscape of several African countries, a communication milieu that facilitates the use of communication and information channels to express diverse views and opinions on national development concerns, particularly in South Africa, Ghana, Mozambique and Tanzania, among others. Qualitatively and quantitatively, much more effort remains to be done in Africa to address the constraints that confront the practice of development communication. However, given the correlation between communication development and development communication, efforts should contribute to an enhanced use of communication and information in socio-economic, political and cultural development processes in Africa.

The effectiveness of participatory development communication: Some contextual factors

A search through the literature on development communication in Africa indicates that a number of factors come into play in determining the effective application of development communication approaches in support of national programmes in the African context. Among these factors are:

- creating a participatory communication environment that not only gives room for the expression of diverse ideas on societal developmental concerns, but also facilitates grassroots-level interaction;
- strengthening the flow of public information and opportunities of public dialogue on development policies and programmes;
- informed popular participation based on enhanced access to pluralistic and independent communication media;
- producing and disseminating information content that reflects as well as responds to the local values and information needs of the people at the grassroots level;
- using culturally appropriate communication approaches and content;
- using community communication-access points, especially community radio and, more recently, community multimedia centres, as well as small-scale, localized and group media;
- ensuring access to information for women and young people and developing their competencies and skills in the use of communication and information technologies;
- harnessing the strengths of traditional media (drama, dance, songs, story-telling, etc.) and combining them with new information and communication technologies; and
- providing practitioners with appropriate training in the use of communication and information to support development programmes.

In the context of development programmes in rural and marginalized communities where the majority of the populations in most African countries reside, community communication-access points, traditional media and culturally appropriate communication approaches and content are of particular importance in participatory development communication. Alumuku and White (2004) have observed that the communicating capacity of the local community must be harnessed in the conception of development communication strategies in the region. In this regard, community media in African countries, especially community radio, provide the enabling space for local community members to make known their views and opinions on development problems and the possibility of participating in the resolution of those problems. Alumuku and White's (2004) study in Ghana, South Africa and Zambia reported that the communicating capacity of the local community in the form of community radio stations was harnessed to produce and disseminate programmes dealing with such issues as healthcare education; conflict resolution; gender equity; education for responsible democratic governance; defending local development interest; stimulating economic development; and promoting local culture. These are symptomatic of the development problems in many African communities, which the power of community communication resources (with their horizontal, participatory approaches) can help to resolve at the grassroots level.

Similarly, in the African communication environment, given the limited access that some national population groups, especially the marginalized segments living in remote villages and rural communities, have to mass communication media, the communicating capacity of the local community resides in the so-called traditional media resources and channels (traditional leaders, drama, concerts, songs, story-telling, puppetry, drumming, dancing, etc.). They serve as reliable channels of news and information gathering, processing and dissemination in many rural communities, and often address local interests and concerns in local languages and cultural contexts which the community members can easily understand and with which they can identify. Effective applications of participatory development communication approaches and strategies at the grassroots and community level should necessarily involve the use and harnessing of these pervasive traditional communication instruments and resources. Traditional media, especially story-telling, songs, drama and local street theatres, stem from local cultural norms and traditions; their content is usually couched in culturally appropriate ways and they often serve as effective means of channelling development issues. They have been used in communication interventions addressing issues related to improving agricultural productivity, natural resources and environmental management, HIV/AIDS and other development problems. Examples of such use abound in African countries.

Illustrative of the use of traditional media to address development challenges are:

- the Theatre for Community Action project in Zimbabwe, which uses theatre to support and involve rural community members in several districts in Matabeleland in the combat against HIV/AIDS;

- the use of theatre and folk musical groups to disseminate agricultural information to farmers in rural communities in Nigeria;
- the transmission of messages about reproductive health and HIV/AIDS prevention through traditional dance and music in rural communities in Ghana, Kenya, Malawi, Mozambique, Tanzania, Uganda and Zambia, among others;
- the use of participatory drama and folk music to address issues of gender inequality and HIV/AIDS in Niger State, Nigeria; and
- the transmission of messages about natural resources management through dramatic performances in Uganda.

In sum, traditional media provide horizontal communication approaches to stimulating discussion and analysis of issues, as well as sensitizing and mobilizing communities for development. However, one must be cautious about romanticizing the abilities and impact of traditional media in development. Like other communication and information means, they have their weaknesses and limitations in time and space; they are particularly deficient in simultaneous dissemination of information about development issues across wide and geographically disperse populations. Research and experience in the use of traditional media indicate that they are most effective in participatory communication of development in rural communities when combined with mass communication resources, especially radio. The challenge facing practitioners of participatory development communication in African countries is to be sufficiently cognizant of the potentials and limitations of traditional media and knowledgeable about how to skilfully harness and combine them with other communication and information forms for development. The practical and technical guidelines for designing and implementing PDC interventions using traditional media approaches in combination with other communication forms (including community needs analysis; designing and pre-testing messages/content; training; costs analysis; raising of required funding; implementation; monitoring; and evaluation) lie outside the scope of this chapter.

Conclusions

Communication and information have significant functions to fulfil in supporting and fostering socio-economic, cultural and political development and transformation in African countries. These functions have been recognized by communication scholars, researchers, trainers and practitioners alike, and constitute the bulk of the literature on communication and development in Africa. They are equally stressed at different levels of communication educational and training programmes and provide the basis for communication practice across the continent. This chapter has attempted to situate the use of participatory development communication within the social, economic, political and cultural development challenges and realities in the region, and has drawn particular attention to a number of contextual factors that determine the effectiveness of PDC in Africa. The chapter has been based on the conceptual premise that the kernel of communication teaching, research and practice in Africa lies in their contribution to addressing the myriad development problems and challenges facing the continent.

Notes

1 The views expressed in this case study are those of the author and do not necessarily reflect those of the United Nations Educational, Scientific and Cultural Organization (UNESCO).
2 See, for example, Jefkins and Ugboajah (1986); Akinfeleye (1988); Boafo (1991); Moemeka (1994) and Kasoma (1994). On training, it is worth noting here, in particular, the introduction of the Masters programme in development communication in the Department of Mass Communication, University of Zambia; the programmes at the Southern Africa Development Community Centre of Communication for Development (SADC-CCD) in Harare, Zimbabwe; the Centre for Rural Radio Development (CIERRO) in Ouagadougou, Burkina Faso; and practical training provided by development institutions, NGOs and community groups. With regard to study materials, one can cite the training modules on development communication prepared by the African Council for Communication Education under a project funded by UNESCO in 1991.
3 For comprehensive discussions and analyses of participatory communication for development, see Servaes et al (1996); Dervin and Huesca (1997); Servaes (1999); Wilkins (2000); Melcote and Steeves (2001); and Huesca (2002).
4 For an analysis of current development problems and challenges facing African countries, see, for example, United Nations General Assembly (1994); OAU (1995); and Secretary-General to the United Nations Security Council (1998).
5 For a discussion of these constraints, see, for example, Boafo and George (1991); Moemeka (1994); and Agunga (1997).

References

Agunga, (1997) *Developing the Third World: A Communication Approach*, Nova Science, Commack, NY

Akinfeleye, R. (1988) *Contemporary Issues in Mass Media for Development and Security*, Unimedia, Lagos

Alumuku, P. and White, R. (2004) 'Community radio for development in Africa', Paper presented in the Participatory Communication Section, 24th General Conference of the International Association for Media and Communication Research, Porto Alegre, Brazil, July 2004

Ascroft, J. and Masilela, S. (1994) 'Participatory decision-making in Third World development', in White, S. A., Nair, K. S. and Ascroft, J. (eds) *Participatory Communication: Working for Change and Development*, Sage Publications, New Delhi

Boafo, S. T. K. (1991) 'Communication technology and dependent development in sub-Saharan Africa' in Suusman, G. and Lent, J. A. (eds) *Transnational Communications: Wiring in the Third World*, Sage Publications, Newbury Park, London and Delhi

Boafo, S. T. K. and George, N. (eds) (1991) *Communication Processes: Alternative Strategies for Development Support*, ACCE, Nairobi

Dervin, B. and Huesca, R.T. (1997) 'Reaching for the communicating in participatory communication: A meta-theoretical analysis', *The Journal of International Communication*, vol 4, no 2, pp46–74

Huesca, R. (2002) 'Participatory approaches to communication for development', in Mody, B. and Gudykunst, W. (eds) *Handbook of International and Intercultural Communication*, Sage Publications, Thousand Oaks, CA

Jefkins, F. and Ugboajah, F. (1986) *Communications in Industrializing Countries*, MacMillan, London

Kasoma, F. (ed) (1994) *Journalism Ethics in Africa*, ACCE, Nairobi

Melcote, S. and Steeves, H. L. (2001) *Communication for Development in the Third World: Theory and Practice for Empowerment*, Sage Publications, London

Moemeka, A. (ed) (1994) *Communicating for Development: A New Pan-Disciplinary Perspective*, SUNY Press, Albany, NY

OAU (Organization of African Unity) (1995) *Relaunching Africa's Economic and Social Development: The Cairo Agenda for Action*, OAU Council of Ministers 17th Ordinary Session, 25–28 March 1995, Document ECM/2 (XVII) Rev. 4

Quebral, N. C. (1971) 'Development communication in the agricultural context', Paper presented at the symposium In Search of Breakthroughs in Agricultural Development, College of Agriculture, University of the Philippines, Laguna, the Philippines

Secretary-General to the United Nations Security Council (1998) *The Causes of Conflict and the Promotion of Durable Peace and Sustainable Development in Africa*, Report of the Secretary-General to the United Nations Security Council, 16 April

Servaes, J. (1999) *Communication for Development: One World, Multiple Cultures*, Hampton Press, Cresskill, NJ

Servaes, J., Jacobson, T. and White, S. A. (eds) (1996) *Participatory Communication for Social Change*, Sage Publications, Thousand Oaks, CA

United Nations General Assembly (1994) *African Common Position on Human and Social Development in Africa*, United Nations General Assembly, Document A/Conf.166/PC/10/Add.1, January

Wilkins, K. (ed) (2000) *Redeveloping Communication for Social Change: Theory, Practice and Power*, Rowman and Littlefield, Boulder, CO

III

Participatory Development Communication in Action

The Old Woman and the Martins: Participatory Communication and Local Knowledge in Mali

N'Golo Diarra

If participatory communication is to bring about lasting change, it must give a prominent place to local knowledge. Yet, such knowledge is not always readily accessible, nor is it easy to deal with the ethical issues that can arise when we are looking for the best ways to put it to use. At that point we must ask ourselves whether the prerogatives associated with our role as communication facilitators give us the right to popularize knowledge that was hitherto the exclusive preserve of a few individuals for whom it may traditionally have been a source of advantage and even of power. The question becomes even

thornier when the knowledge in question could turn out to be very useful for the community as a whole.

During my field research for preparing a training kit[1] on ways to combat erosion, I came across a piece of indigenous knowledge that for decades has been improving agricultural output and human welfare in a village tucked away in Mali's cotton zone, more than 500km from the capital city of Bamako.

I was researching local farmers' knowledge as well as new, supplementary techniques that could be used for combating farmland erosion. My strategy was to hold meetings with various social and occupational groups in the village according to an agreed schedule that would fit with their activities.

In these discussions, farmers showed that they knew a lot about their environment and about the best ways to preserve it. They told me, moreover, that fighting erosion would mean curbing the wholesale cutting of trees on the plateaus surrounding the village. But they also pointed out that, because their fields are on hilly ground, they would have to align the furrows to counter the flow of rainfall runoff. Some of them even suggested that fallowing could be a solution; but most participants eventually discarded this idea because the extensive nature of local agriculture left little room for fallow fields.

After three days of investigation, I went to brief the village chief on my findings before reporting them to a public feedback session. I was very happy with the cooperation the villagers had shown and with the results I had obtained, and I was thinking only about getting on with developing my training kit. So I was surprised when the village chief put a question to me, seemingly out of the blue: 'Has anyone told you about the governess of the seasons?'

I had to confess that I knew nothing of her. But I had no hesitation in countering with another question by asking him who she was. Smiling at me in a kindly way, he tapped my shoulder and invited me to supper. The next day, very early in the morning, the chief woke me and without taking time for breakfast, he took me to the house of the old woman who greeted us as if she had been expecting our arrival. The chief made the introductions and explained briefly to the old lady what I was doing in the village. Then, excusing himself, he disappeared and left me alone with the 'governess of the seasons'.

She began by recounting to me something of her own life, including the tragic death of her husband and the loss of her six children. It was a lonely life, she said, but yet she felt it was a happy one. Suddenly, she went into a trance and began to speak in words that I could not understand. Somewhat disconcerted, I managed nevertheless to keep my cool. She soon got hold of herself, looked at me a moment and then asked me if the 'purple martins' had arrived. Surprised by this question, I had to say I had not noticed them. At that, she burst out laughing. She asked me to sit down beside her on the couch and began to tell me why she was known in the village as 'the governess of the seasons'.

'My father was the great traditional healer in the village', she said very matter of factly:

When he died, he would have no direct descendents he could pass on his knowledge to, except me. One day he called me to his hut. He said that he was at the twilight of his life, that there were things he knew and that he must pass on to his descendents. But he could not transmit everything to me because I was a woman married into another family. I told my father that whatever he could pass on to me would be of service to the whole village, for I no longer had any children in this world to whom I could transmit that knowledge. It was then that my father told me the story of the purple martins and how they govern the local seasons.

Looking me straight in the eye, she continued her story:

'Remember this well, daughter', said my father, 'the purple martins are fabulous birds. They arrive at the start of the rainy season and they make their nests in the forest along the great river that passes by the village. Their arrival, and that of the storks, heralds the beginning of the rainy season for people in our village and nearby.

When they build their nests down in the river valley, I have noticed that they take into account the volume of water expected in the river. Three weeks after their arrival, I find that their nests will be placed either high up in the trees or lower down. So I conclude that, since they are nesting when water levels are high, the position they choose for their nests will depend on how much water they expect the river to have, perhaps to prevent their eggs from being washed away. As soon as I have noted the positioning of their nests, I consult the spirits and my fetishes for permission that very evening before I tell the news to the village chief so he can order sacrifices. The announcement of this news is a big event for the entire village. To some extent, it determines what farmers will plant and where they will pasture their livestock during the coming season. All the families promise gifts if they get a good harvest and if the village workforce stays in good health.'

Continuing her story, the old woman confided to me that her father was now dead and that she continues to play this role of village seer. 'That has been the source of wealth for my family and for the village for generations, for after the harvest all the villagers keep their promises, and our village and the neighbouring area are assured of food self-sufficiency.' She concluded with a laugh: 'There you are, thanks be to the purple martins!'

With the old lady's concurrence, and my commitment not to reveal her name or the name of the village, I have been able to use this example as part of the scenario for the training package.

The community sees things differently

Two years after I heard the old woman's story, I had the chance to return to the same village for a farmer training session using the anti-erosion training package we had produced at the time. There were warm greetings all around and the occasion was marred for me only by learning that the village chief and the old woman had both died.

The training began well. As agreed, I used the example of the old woman and the martins, treating it as if the story had taken place somewhere else. The villagers were all for trying to protect these birds by taking better care of the riverbed. Yet, when I applied the example to the river that runs past the village where we were, there was a chorus of outcries from all sides and the training session was temporarily disrupted.

Some farmers went so far as to declare that this river was haunted by evil spirits, or djinns,[2] to whose wrath could be laid all the flooding that has afflicted the village and its surroundings. No one talked about how important the river was for the village's social and economic development and still less about the fact that after the rain all the water runs off the plateaus straight into the river. In the villagers' minds, the mystic dimension of the river far outweighed its role in terms of the use and conservation of local resources.

Local knowledge and ethical issues

Local knowledge is very difficult to get a handle on. It is usually held by individuals or families of great status and influence in the community, as was the case in this village. Because of this, we participatory communication practitioners must weigh the pros and cons before we disclose any such knowledge. I found myself in a very tricky situation that day and it left me hesitant for a long time about the wisdom of revealing the old lady's story. I finally decided to talk it over first with the new village chief. After some consultation, he gave me the go-ahead to tell the tale to participants in the training session.

When I started my story, there was dead quiet. Everyone was surprised at what they heard. Feelings of guilt, astonishment, curiosity and even disgust registered on their faces and in their statements. A few were against my revealing this secret of such importance to the community. Most of the participants, by contrast, were relieved to hear the story. Many of them felt guilty about their practices or about their behaviour towards me, as the trainer, but above all about their treatment of the river – this resource that was so vital to the community but that had been blamed by so many generations for the flooding it caused.

In this way, a training course in anti-erosion techniques to preserve farmland was transformed into a real discussion about changing the community's attitude towards the river. All the debate focused on how to save and protect the river. At the end of the session, we addressed the need to adopt measures to protect the river from erosion in order to avoid flooding in the villages. Then we went on to debate the importance of restoring farmland to ensure the community's economic development and its food security. Participants were unanimous on the need to take anti-erosion measures to protect the river and make the villages safe from flooding. To do this, they set the goal of creating a buffer zone along the river, which would also serve to protect the martins. They also undertook to use anti-erosion techniques on the plateaus in order to break the speed of the rivulets and the wind so that the fields could be restored and agricultural output improved.

All these commitments are very fine; but now that this secret is out, what will come of the vision and of people's attitude towards the possessors of local knowledge? What will become of places that used to be considered sacred? As development practitioners, how can we make judicious use of local knowledge, while respecting ethical, copyright and intellectual property issues?

Drawing lessons from the experience

One of the lessons I drew from this participatory communication experience was that the villagers were very happy to have me back and especially to see themselves in the film. They were greatly bolstered by the opportunity to express themselves and to be heard. They realized that they had something to say about everything that concerns their local development and they lost no time in letting me know that.

It is clear that, thanks to participatory communication, these farmers have been able to reach consensus on the key role that the river plays in their community and on the best ways of managing that resource. In the past, the old lady had been regarded in the village as merely an oracle who had to be consulted at the onset of the rainy season so that the good farmers could set their course. The martins were looked upon by many people as simple migratory birds whose arrival announced the beginning of the rainy season, while the river was considered a place haunted by evil spirits. Indeed, if the old lady's story had not been revealed, it would have been much harder to get people to protect the watercourse – they would have continued to treat it as the haunt of *djinns*, and it would have been left to its sorry fate. In that case, neither the farmers' fields nor the river would be safe today.

The achievements of participatory communication

In the experience recounted above, it is clear that new know-how for combating erosion, coupled with indigenous knowledge, including that of the old lady, led to a more rational and efficient approach to managing the community's natural resources – that is, the river and its surroundings, farmland, pastureland and the village itself, which was always subject to flooding from rain runoff.

Another important point is that this village has now become a pioneer in combating erosion through such techniques as planting hedges or 'green fences', building stone dykes and weirs or planting sodded strips, and it has even enlisted neighbouring villages in the cause. Farmers have put up stone-retaining walls all along the river and on the hillsides surrounding the village in order to break the speed of rain runoff. This has helped to protect the river and, at the same time, to make the hilly fields usable for dryland farming, while reserving the lower-lying lands for rice growing.

Villagers have also planted trees in a strip about 300m wide along each bank of the river. They have banned all woodcutting and fires in this zone, except with explicit permission from the village chief and his advisers, who make up the village's forest management committee.

Many activities are now flourishing along the river, including fishing, livestock raising, rice growing and, in particular, market gardening, to which the women devote themselves for a good part of the year. The village now has its own little forest along the embankments where the martins can come to lay their eggs every year without fear. The area surrounding the river has today been transformed into a tropical microclimate where people, animals and birds together share the bounty of the legacy left by the old woman and her martins.

Of course, there is still the question of sustainability for these achievements, particularly since the village chief and the old woman are dead. Yet, all signs suggest that farmers' involvement in efforts to protect the river will help to keep these actions going.

By way of conclusion, this experience has made clear to me that a community's local knowledge must have an important place in the entire process of participatory communication. It is up to us, communicators, to learn how to discover that knowledge, to be in phase with its holders and to introduce it into our training tools so that communities can put it to use in the interest of local development.

For the village chief and the 'governess of the seasons' who taught me to appreciate and to pass on this knowledge, my prayer is that they may rest in peace.

Notes

1   A training kit is a set of audiovisual and hardcopy training tools prepared in cooperation with the community. The process, as developed by the Centre de Services de Production Audiovisuelle (CESPA) in Mali, involves several stages during which we structure the contents in light of local people's knowledge about a topic, taking due account of their cultural codes and of the objectives for the training exercise. The outcome is, thus, the result of a joint effort among researchers, development technicians, communicators and rural people themselves.

2   In the cosmology of some Malian cultures, the *djinns* are evil spirits who usually dwell near watercourses. Natural disasters are often attributed to the wrath of the *djinns*, when they have been displeased by some human action.

Introducing Participatory Development Communication within Existing Initiatives: A Case from Egypt

Introducing participatory development communication (PDC) within existing initiatives is not an easy task. The necessary conditions must either be in place or put in place. Attitudes must also reflect the methodology that one is trying to introduce. When these are lacking, collaboration between the different stakeholders can be impaired. This is particularly true when the various partners work on different levels and weigh differently in the

decision-making process. This chapter presents the challenges faced when getting on board after an initiative has started and when trying to introduce a methodology that questions the way things have been done so far.

In Egypt, 95 per cent of the population is concentrated in 5 per cent of the entire country's surface area. Therefore, the government of Egypt has asserted that developing human settlements outside the Nile Valley in desert areas and frontier governorates is a national priority.

However, the expansion of human settlements outside the old valley faces several constraints, such as the inhospitable and vague institutional set-ups for managing development in desert areas, outside current jurisdictions. The poor coordination and cooperation between the existing governorates' local authorities and new communities is also a problem.

In that context, the government office in charge of urban planning has been assigned the responsibility of preparing regional, urban and structural plans for the new settlements. It is also responsible for monitoring their implementation. Hence, close working relationships with local authorities have to be in place, especially since the technical capacity of local governments does not yet allow for comprehensive decentralization.

With a view to responding to these difficulties, an initiative was put in place with funding from both the government office in charge of urban planning and the United Nations Development Programme (UNDP). Its main objective was to devise implementation mechanisms for strategic plans in order to create an enabling regulatory and procedural environment for sustainable development and growth. The initiative was to start with a pilot demonstration in two relatively new communities.

More specifically, the initiative aimed at improving the capacities of the government's urban planning office and of local partners in participatory planning and management, in which participatory development communication plays an essential role. It also aimed at helping the initiative's partners to institutionalize these new capacities, establish partnerships and create strategic alliances to attract investments and sustain implementation. These partners included the government's urban planning office, the local authorities of the two communities involved, small- and medium-sized entrepreneurs, business associations and the communities at large. The preparation and dissemination of guidelines for city and community consultations were also considered as part of this initiative, as well as holding consultations to prepare local development plans in light of the strategic plan.

Origins of the initiative

This initiative was born after three years of awareness building and sensitization efforts led by an institutional development expert on the importance and benefits of participatory approaches in development planning. Early diagnosis had shown that the participatory planning and management practices of the human settlements in the area under issue could greatly benefit from being expanded. It also showed that the initiative's partners needed assistance in adopting a more inclusive *modus operandi*, with a view to establishing partnerships and strategic alliances that would gear the strategic plan towards its implementation.

In the course of that preparation phase, it became clear that participatory development planning is much broader than the strategies strictly developed from a physical planning viewpoint, which have proved to be a complete failure. Indeed, their failure has been manifest in the fact that although new cities were designed and plans were handed down to local governments for implementation, the strategies were never fully implemented. Therefore, citizens were reluctant to move to these new cities. Those who did move faced huge difficulties as the services provided did not respond to their needs or were absent in the first place.

From theory to action

In its early phase, the initiative worked towards streamlining the inter-organizational interface within the existing institutional framework. It tried to detect the most serious barriers that needed to be reconsidered within that framework in the area under issue, and proposed realistic modifications in order to enhance the potential of public, private and civil society participation in local development.

Thus, through direct and constructive dialogue and consultations, the initiative tried to help stakeholders to define and rectify sources of friction and derailments, and proposed practical institutional changes. In addition, municipal policy options paved the ground for an improved business environment and participatory development.

In order to introduce participatory development communication within the initiative, several steps were envisioned. First, all the documents related to the initiative were to be reviewed in order to become aware, for example, of its nature, the main problems encountered, suggested solutions and the stakeholders involved.

The second step consisted of assessing and analysing the communication problems, as well as the communication channels and materials used by the immediate stakeholders.

Third, a training programme was to be developed and implemented in order to help the initiative's partners understand the concept of participatory development communication and acquire the capacity to apply it in their work. In order to do that, it was necessary to identify a unit within the existing organizational structure of the communities' local authorities that could become the key actor in maintaining the use of participatory approaches throughout the implementation of their communities' development strategies. Thus, the idea was for the staff of these units to be trained in participatory communication and development planning, together with staff members of the government planning office.

The training was envisioned to be of a practical nature and to emanate from the specific circumstances in which the participants live and work.

Hence, it was expected to facilitate the application of their newly acquired skills directly in the field. Initially, the idea was for the staff of these units to be trained in participatory communication and development planning so that they could implement their communities' development strategies. The training programme was to be held over a lengthy period of time in order to ensure that participants actually learn and practise the skills needed for a participatory development communication process.

The next step involved providing assistance to the immediate stakeholders in developing a participatory communication strategy for their communities' development plans. Finally, the plan included participation in general activities such as workshops that lead to the development of such elements as issue papers and city consultations.

Results fall short of expectations

Several obstacles hindered the implementation of the plan described above. First, the project team within the government's planning office continued to perceive participation as a process aimed at allowing all stakeholders to voice their problems, but without helping them to overcome these problems.

Furthermore, the team significantly interfered in the work of the communication and development planning specialists, with a disruptive effect. It did not allocate enough time to assess the training needs prior to organizing the training programme. It also continuously requested pre-prepared training material, as if distributing bulky materials to trainees would ensure the training's success.

Probably the greatest difficulty was that the team viewed training as a top-down process in which lecturing was to be the main training tool. In the eyes of team members, the training duration was not to exceed two to three days (for budgetary constraints). Moreover, these short training workshops included both communication and development planning. As a result, there was not enough time for participants to fully grasp what participatory development communication really means and where it fits within the development planning cycle. The same situation prevailed with regard to the institutionalization of participatory approaches in the government's planning office and the local authorities' structures: the resources allocated did not suffice in terms of time and money.

Finally, the staff of the special units had not yet been fully selected at the onset of the programme and there were many disagreements on the selection process (and too much politics was involved). Thus, the implementation of the training sessions was disrupted, with a significant impact on the participating structures' capacity to integrate participatory methodologies in a sustainable manner.

Looking back on the initiative

Despite its serious setbacks, this initiative provided lessons on introducing participatory development communication within existing initiatives, particularly if more than one funding partner is involved. In hindsight, it seems that several preparatory steps would have helped the initiative to better achieve its goals.

For one, the international organization supporting the initiative has to be aware of participatory development communication, its benefits and prerequisites. This awareness would facilitate the work of communication specialists in many ways. For example, the terms of reference would be formulated more adequately and the communication facilitator would find support when it comes to dealing with the governmental partner. Unfortunately, up to now, many donor agencies or international organizations still do not fully understand the principles of participatory development communication. Many of them seem to be more concerned with the rights-based approach and making efforts to introduce it within all their programmes. Participatory development communication ought to be treated in the same way.

Furthermore, participatory development communication should not be considered a 'complementary accessory' to the initiative to be introduced at a later stage (mainly the implementation stage), as was the case in the initiative described in this chapter. Participatory development communication should start with the idea of the initiative and this, of course, is only possible when the partner international organization is fully aware and convinced of its value.

Accordingly, the governmental partner should also be trained in participatory development communication's concepts and methodology. This training would enable the communication specialist to work hand in hand with the governmental partner to develop a participatory development communication strategy for the initiative that they are about to implement. If all these conditions are in place, one can expect that the governmental partner will support the PDC process throughout the project instead of hindering it.

Goats, Cherry Trees and Videotapes: Participatory Development Communication for Natural Resource Management in Semi-Arid Lebanon

Shadi Hamadeh, Mona Haidar, Rami Zurayk,
Michelle Obeid and Corinne Dick

When societies undergo in-depth changes, traditional structures, resource management systems and means of communication sometimes do not suffice in coping with the pace of transformation. Moreover, conflict may arise as a result of the increasing stress that people and their environment are faced

with. Enhancing existing communication and conflict resolution practices with novel tools and methodologies can help to build the social fabric anew and provide communities with a common sense of direction.

Traditional agro-pastoral Lebanese villages located on the marginal slopes of the semi-arid Lebanon mountains have been undergoing drastic changes in response to socio-economic pressures that have been developing over the last 25 years. During 1991 to 1993, a case study on changes in resource management systems conducted in Arsaal, a vast highland village, revealed a massive conversion from a traditional cereal/livestock-based economy to a rain-fed fruit production system.

Subsequently, a follow-up study was conducted with a view to analysing components of changes, trends and sustainability in the emerging production system and improving the prospects for sustainable community development. The land-use system in Arsaal, including socio-economic components, was then characterized and its resource base was assessed, with an emphasis on soil and water conservation strategies. Local beneficiaries were involved at different stages of the project and strengthening of local capacities was sought through the establishment of a local users' network. Avenues for non-agricultural income-generating activities for women were explored. A second phase was designed to test and evaluate technologies and management options developed by the users' network in phase 1 to assess progress towards sustainability in the major land-use systems, as well as to monitor, evaluate and strengthen the capacity of the local users' network towards sustainable natural resource management, with an emphasis on gender analysis.

The overall picture reflected a society in transition from an agro-pastoral system to a more diversified livelihood integrating rain-fed fruit production and off-farm jobs, as well as quarrying and related activities, in addition to the traditional sheep and goat and cereal/pulses production. Traditional management strategies based on community consensus had given way to conflict over land use among animal herders, fruit growers and quarry owners, with an increasing socio-political influence from the latter group. The conflict was highly entangled in the traditional family web of the village. In spite of the ramifications that these conflicts introduced within families and clans, traditional structures showed a good degree of resilience. In summary, the Arsaali society was in a prolonged state of disequilibrium, crisis and crisis management.

Furthermore, starting in the mid 1960s, the breakdown of traditional resource management practices had led to the dismantling and complete paralysis of the local municipality. This situation was perpetuated during the decades of civil unrest because the new emergent forces (political parties and militia) were more involved in national politics than in local resource management. This chaotic state of affairs continued following the return of hundreds of youth from Beirut, driven back by the demise of the leftist militia after the Israeli invasion of Lebanon and their hopes of prospective involvement in smuggling activities across the Syrian borders.

The year of 1998 saw the election of the first municipal council in 35 years, and the community acquired an administrative body that faced the challenge of managing several forms of land use and their conflicting requirements. The election, however, was largely thought of and conducted in terms of familial alignments, thus leading to the formation of a municipal council that knew little about local administration and lacked the perception required to develop local resources.

Creating a local users' network

Following discussions with our local facilitator, the Association for Rural Development in Arsaal (ARDA), it became crystal clear that some kind of medium was needed to facilitate interaction among the various local beneficiaries and other groups, such as researchers, development workers and non-governmental organizations (NGOs). The presence of such a medium would provide a platform for different stakeholders to assess and develop a common understanding of research and development needs and possible solutions to the lack of extension structures.

A local users' network was therefore conceived to bring together the different stakeholders and to fulfil the critical functions of participation, communication and capacity-building.

In the Arab world, the traditional way of communicating and resolving dilemmas is, largely, face-to-face interaction. This forms the basis of the tribal *majlis* during which issues are brought up in the community, usually at the house of the community leader.

The strategic role intended for the network was to form a participatory interactive platform based primarily on face-to-face interaction in informal group meetings as a variant of the traditional tribal *majlis*, this time extending beyond the community and involving all the development stakeholders: the community, researchers, development projects and government.

After consulting with the local NGOs and meeting with the local authorities, the *mukhtars* (heads of villages) and the acting municipality officials (the municipality had been dissolved since 1965), the mechanics for establishing the network were defined in order to ensure the representation of traditional decision-makers, as well as new emerging forces. ARDA played a facilitating role in contacting various groups of users (such as the cherry growers, flock owners and women).

The objectives of the project were discussed and evaluated, and network members agreed to actively participate in the activities.

It was hoped that the flexible structure of the network, the common interest of its members and the rewards to all involved in it would ensure a level of sustainability through its adaptive development. The rewards included benefits arising from farmer-to-farmer training, productivity improvement as a result of scientific research, improved training capacity of the NGOs involved, and reinforced links between farmers and local authorities.

As the network grew, our understanding of communication principles evolved with it and the need to define a workable, meaningful typology or a system of user categorization that considered their subjective nature became obvious. Specialized working groups were born and later developed into three sub-networks; two of these dealt with the main production sectors in the village: livestock and fruit growing. The third addressed women's needs. Local coordinators were designated to coordinate each sub-network. Specific on-farm trials were developed, discussed with the farmers and implemented. From its onset, the project made sure that network members were representative of the different resource user groups in the community. This helped to ensure that the needs of the community at large were voiced in the network, which meant that the solutions developed were relevant to the rest of the community. This, by itself, greatly enhanced a widespread knowledge-sharing.

Moreover, a specialized unit in the network called the environmental forum was specifically created to catalyse knowledge-sharing with the community at large. The forum was made up of Arsaali youth, mostly school teachers, who were trained by the project team in the good practices developed by the network and had as a mission their widespread distribution. In order to do this, they primarily used face-to-face interaction, especially during critical periods. In addition, they used complementary material, such as the 'best practices booklet' developed by the project to summarize and simplify project findings in a language accessible to farmers. The forum also served as a communication channel between the community and the local users' network, in which refinements and remedial measures were identified.

Special emphasis was placed on evaluating and analysing the observations and feedback of the network members. This was done with the purpose of assessing the response of the community to the new techniques, as well as adapting these techniques to community needs. Farmers' findings were fed into the research process by way of regular meetings and contacts with research team members. These farmers constituted a platform whose purpose was to spread research findings and exchange observations. The input and feedback of network members constituted the main elements for use in establishing intervention strategies to gain sustainable improvement.

Using a wide range of tools

The tools and practices used by the local users' network were mainly interpersonal. They included regular issue-centred, round-table meetings for members of the sub-networks; community outreach by students during their training programmes; 'live-in-the-village' and 'work-with-the-farmer' approaches; joint field implementations of good practices in natural resource management; short video documentaries on different issues, which were also used as powerful participatory tools; newsletters; a website; and, most importantly, a series of workshops on different themes related to natural resource management and community development.

The network functioned as a self-reinforcing interactive participatory communication platform that proved to be an effective and innovative experience by promoting economic development and socio-political empowerment, while exposing the community to other development interventions.

Art and visual communication tools were also part of the network's arsenal. In its second phase, the project developed a partnership with Zico House, an alternative cultural community house specialized in the use of art for community development. Video-making was experimented with as part of an effort to involve the community in dialogue and conflict resolution, the premise being that imagery has the power to shed light on aspects of conflict and dissent. More importantly, it constitutes a platform for freedom of expression for marginal groups and provides a visual reference of a specific development context over time.

Making videos to resolve conflicts

The network provided an environment in which conflict resolution could take place among different land users since the needs of conflicting parties could be voiced and compromises explored. Early meetings held to discuss the conflict issues among the parties revealed a reluctance to engage in dialogue. After a few stalling sessions, our communication unit suggested using visual images to facilitate the dialogue initiation process. Initially, the representatives of the conflicting parties who refused to discuss the issues at stake were interviewed and filmed separately. Then the video was projected in the presence of all parties, followed by a discussion that was also documented on film. The local actors who were videotaped were consulted during the video editing process. The final step was to project the new video to a larger audience, including more people from the whole village until a positive dialogue started to emerge from the audience. The moment of consensus was also filmed and formulated by a local facilitator into a set of specific recommendations for follow-up. For instance, the conflicting parties agreed to refer to the local authorities (the municipality and the *mukhtars*) and entrust them to develop and recommend different scenarios for land-use management in the village.

Videos to empower women

Another video was produced with the aim of highlighting the economic productivity of women in NGOs, in the co-operative for food provisions and in a pastoral society. This video particularly addressed the improved self-perception of women and the feeling of empowerment that accompanies production. The film was shown to a group of women and men, some of whom were in the film, over an *iftar* (breakfast) organized by ARDA. Those who were in the film felt quite empowered to see themselves on the screen, especially when they received compliments from others about their stated opinions in the film. Both men and women emphasized the importance of working extra-domestically. Women stressed that although money is essential and it elevates the status of women in their households, the mere act of exposure, learning and socialization that comes about from working is satisfying and, indeed, raises one's self-esteem.

There was also a discussion about women who are not involved in production and the importance of acknowledging their role in society. Some commented that the film should have also addressed other 'typologies' of women, such as school teachers and housewives.

The group agreed that it enjoyed watching the film and that film is a very appropriate means of documentation, especially in a context such as Arsaal where people enjoy 'watching' more than reading.

Conclusions

The use of a wide range of communication tools and methods – that is, from the traditional *majlis* setting, to workshops, to novel tools such as videos and video-making – in the local context was very revealing. Marginal groups usually shut out of the local power structure, suspicious of it and often shy in formalized *majlis* setting became very candid before the camera and expressed unvarnished opinions as if, for them, lenses were neutral objects and there was no need for the formal politeness of the *majlis* confrontation. The videos turned out to be valuable for generating discussions and awareness among and between different people and factions. The world of image was able to re-establish a communication platform to implicate local people, reflect their real needs, allow marginal groups free expression, shed light on the nature of conflicts amidst the village, and facilitate resolution of conflicts over natural resource management.

One of the most pressing issues we still face is how far the local communities will be able to use the product of participatory development communication to improve their livelihoods. Participatory development communication activities must be intimately linked to development activities – namely, the transfer of resources. Only when elements of development were injected into our community-based research process did changes in the behaviour and aspirations of the people start to emerge.

From Resource-Poor Users to Natural Resource Managers: A Case from West Java

Amri Jahi

Conservation efforts often fail to take into consideration that alternatives have to be available if local populations are to change their natural resource utilization patterns. Along the Cimanuk River in West Java, efforts to prohibit agriculture on the raised riverbanks to curb seasonal flooding irremediably failed until viable economic alternatives were introduced. Through participatory development communication, the dialogue between researchers, decision-makers and small and landless farmers made it possible for sheep raising to flourish as an alternative to the practice of agriculture on the riverbanks that was making villages more vulnerable to annual flooding.

Fifteen years ago, as we were preparing a baseline study to explore the development options that could improve the living conditions of small farmers and landless farm labourers living along the Cimanuk River in the district of Majalengka, West Java, we wondered whether those resource-poor farmers would be able to participate in managing a strip of public land stretched out along the river. We also wondered whether they would be able to derive benefits from such an undertaking.

At that time, the main concern of the local government and of the communities living along the river was flood control. Every rainy season, heavy downpours filled up the river and flooded the villages. To control this annual flooding, the Department of Public Works bought a strip of land along the river that goes through the district's villages and small towns, and raised the riverbanks to a certain height. Hence, the annual flooding was controlled and the communities were freed from the yearly threat of natural disaster.

However, this flood prevention technique created another problem for the small and landless farmers who had been farming on the riversides long before the banks were raised. According to the new government rules and regulations, farming activities were forbidden on the public land along the river due to the risks of destabilization of the land structure and hindrance of the waterways.

In spite of the government rules and regulations, the small and landless farmers pursued their farming activities as usual. On their part, officials of the Department of Public Works continued to enforce the regulations by eradicating the crops that could possibly weaken the raised riverbank's land structure.

Looking for alternatives

After several years of playing cat and mouse and tired of conflicting with the local communities, the Department of Public Works finally proposed a win–win solution to the farmers. The latter were finally allowed to farm along the riversides, provided that they grew grass at least 1m away from the river's brim and on both sides of the raised bank. However, they were not allowed to grow bananas or any other type of land crops that could destabilize the land structure.

To further motivate the farmers to grow grass, the government officials promised to supply them with sheep. However, this promise did not materialize very well. The farmers were disappointed with the few low-quality sheep delivered to them and the abundant grass was wasted.

At that critical time, a baseline study was undertaken in order to understand the availability of local natural resources such as land, crops, animals and water, as well as the limitations, problems and opportunities associated with their use and development. Moreover, the study also aimed at understanding the traditional systems and at assessing indigenous capacity.

The study concluded that there was potential for involving the small farmers in managing the public land located along the river and the irrigation channels, and for deriving benefits from it, provided that a funding body be put in place.

A follow-up pilot study was then undertaken. Essentially, the challenge consisted in helping farmers to solve their problems and benefit from the abundant grass, while ensuring the conservation of the riverbank. After discussing the best-suited approach, our team decided to use participatory development communication. Thus, linkages were established between university researchers, local government, the Department of Public Works, the local livestock extension service, the village authorities and local farm communities.

Using appropriate communication tools

Prior to the fieldwork, communication materials such as sound-slide shows, posters and even a comic book were prepared and tested with farmers and extension workers. Development communication graduate students were involved in producing and testing these materials, either through their message design course assignments or through their theses research.

Topics such as the social and economic potential of sheep raising in the district of Majalengka were specifically addressed with local decision-makers and livestock extension workers. Other topics concerning various technical aspects of sheep production, marketing and rural family budget management were developed for extension workers, farmers and their wives.

The fieldwork started once the communication materials were ready. In the first quarter of 1992, a memorandum of understanding was officially signed between the government of the district of Majalengka and the Faculty of Animal Science at Bogor Agricultural University.

During the ceremony, a sound-slide show regarding the social and economic potentials of sheep raising in the district of Majalengka was shown to the government officials and to the local House of Representatives, prior to signing the agreement. This created enthusiasm among the officials attending the ceremony. Moreover, the local media covered the event, which was broadcast on the evening news by the regional television station.

The next step consisted of promoting capacity-building for the local livestock extension personnel and farmer leaders. Once a month, researchers conducted an on-site training meeting for the livestock extension workers and other officials of the district livestock service. The technical aspects of sheep production, including housing, feeding, breeding and sheep healthcare, were addressed. Two sound-slide programmes about sheep production were shown and discussed with the extension workers.

A pre-test and a post-test on the training subjects were always administered to the extension workers prior to and after the slide show in order to build up the agents' interest and to focus their attention on the subjects. Moreover, the tests were also useful in indicating whether the sound-slide shows had been effective in improving the agents' knowledge on the topics that were being discussed.

Following the agents' training, a short meeting was conducted for farmer leaders of the four pilot sites. On that occasion, they were informed that they were entitled to receiving an in-kind loan in the form of sheep. If they agreed, they would have to return a certain number of the sheep offspring to the project for a period of four years.

After the meeting, participants were invited to visit a senior farmer leader in the village of Balida to witness how he collected forage. One month later, they joined an excursion to the district of Garut to visit a sheep-raising initiative. The Garut sheep were particularly well known for their fast growth and prolificacy.

Trying out new techniques

Upon their return, every farmer leader was asked to build a good sheep house capable of housing two adult rams and ten ewes, and to prepare a plot to grow improved varieties of grass and legumes to feed the sheep for demonstration purposes.

Intensive supervision and backstopping activities were provided to the farmer leaders in the following month to allow them to get first-hand experience in raising the new sheep breed in their own environment. After one month, the farmer leaders had enough experience in taking care of the sheep and were able to provide information about the Garut sheep raising to other farmers in the villages.

Upon completion of the trial run among the farmer leaders, monthly evening training meetings were conducted for the rest of the farmers in the four research sites. The first training meeting was about building a good low-cost and healthy sheep house using local materials, while the second one focused on the appropriate feeding for this specific breed of sheep. During the meetings, all farmers were given pre-tests and post-tests in order to measure their knowledge regarding the training subjects prior to and after the audio-visual presentations.

About two weeks after the training sessions, the farmer leaders were asked to check whether all the farmers had made or had improved their sheep houses for the incoming new sheep. Farmers who had not done so were encouraged to rebuild or improve their sheep houses.

Two weeks after completing the necessary preparations, a farmer leader with extensive experience in sheep breeding was asked to join the researchers to select rams and ewes for 100 farmers in the four research sites.

The training meetings continued after the sheep had been distributed to the farmers. The topics addressed ranged from technical aspects of sheep raising to rural family budget management.

Meanwhile, the research assistant, together with the farmer leaders, visited every farmer once a week to check on the sheep's condition and health and to provide further backstopping activities to the farmers.

About one year after the distribution of the sheep, all farmer leaders were invited to a meeting in the district capital to share their experience in Garut sheep raising with farmers from other villages. This farmer-to-farmer communication turned out to be an excellent medium for raising other farmers' interest in Garut sheep production, better than anything that researchers or extension workers could have done.

Promising results

The initiative had a first positive result when the head of the local livestock service, using the narrative progress report for the first year, was able to convince the local government to provide his office with extra funding

Towards the end of the second year, results began to be experienced at the grassroots level, when several farmers began to return the sheep offspring to the initiative. According to a previous agreement, the sheep offspring were to be split fifty–fifty. The returned offspring were to be revolved to other farmers interested in sheep raising. This was an indication that the initiative was having a positive impact on the small farmers involved. Indeed, between September and December 1993, the sheep population increased by 159 per cent. One year later, the increase reached 271 per cent, while at the end of 1995, it had risen to 306 per cent.

The value of the participating farmers' assets also increased accordingly. In less than three years, the value of the sheep almost tripled – a positive contribution to the local economy. Moreover, the welfare of most farmers involved in the initiative improved significantly since the income generated by selling the sheep was worth more than 1.5 times the capital loaned to them.

In addition to the economic benefits, such as having more sheep in barns and gaining additional income, the initiative also brought about some social benefits to the villagers involved, who gained respect from their families and communities. One of the farmer leaders even managed to send his son to the Bogor Agricultural University, where he obtained his degree in Animal Science last year.

In May 1996, after about five years of continued fieldwork, the initiative officially came to an end. However, since the ewes continued to give birth and farmers kept on returning the sheep offspring, the revolving activities continued on their own.

A new spur to the initiative

In early 1999, following the severe Asian financial crises that badly hurt the Indonesian economy, the initiative obtained additional funding to buy 55 new sheep. These sheep were distributed to two other villages in the district of Majalengka.

Meanwhile, around the same time, one former project group in the village of Kadipaten received a 300 million Indonesian rupiah loan through the World Bank's social security network funds. The purpose of this loan was to help the local communities cope with the economic crisis. The farmer group used it to further develop their sheep raising ventures.

Eight years after the initiative terminated, we are still serving the farmers, though at a slower pace and on a smaller scale. To cover the operational costs of supervision and backstopping activities, fieldworkers have been allowed to sell the culled sheep.

Today, livestock production, including sheep production, is growing at a good pace in the district of Majalengka. Both public and private investments have been increasing steadily, especially since the 1998 economic crisis. Between 2002 and 2004, the government of Majalengka invested over 9 million rupiahs in livestock production.

Public funds have been channelled to farmers as loans through local banks, mainly for beef and cattle production. In addition, a big livestock marketing infrastructure has recently been built in the region. Livestock traders from east and central Java bring and sell their sheep and cattle in the newly operated livestock market to local and other traders and buyers from Bandung, Bogor and Jakarta.

In other words, the situation of livestock economics in the district of Majalengka has greatly evolved compared to the situation that prevailed in 1989, when the research team first arrived. We believe that over the last 15 years, our work has contributed to this situation by raising a flag to both public and private audiences, informing them about the existing potential in a remote district of West Java that was waiting to be explored and developed.

During recent times, when driving along the river and the irrigation channels in the former research sites, one can see that the irrigation channels are well maintained and that farmers periodically harvest the grass on the riverbanks. From a distance, one can see numerous small white plastic flags stuck orderly in the ground every 10m, on both sides of the irrigation channel. What do these flags mean? Are they a border line? Yes, they are. No one is now allowed to harvest the grass except for the farmer who sticks the flags on the ground. The grass now means something to the rural communities as it has played a key role in local economic development, while contributing to solving environmental problems.

In summary, through participatory development communication, researchers, extension workers, government officials and cooperating small farmers were able to work hand in hand to solve certain community problems.

Through the introduction of sheep raising as an alternative economic activity, the small and landless farmers living in the farming communities along the sides of the Cimanuk River have been able to ensure the management of the meagre public natural resources in their environment.

Problems encountered

This story would not be complete without describing some of the problems that the researchers encountered in the course of the initiative.

The first problem occurred during the early phase of the fieldwork, when we were trying to gain acceptance and support from the village heads. Most of the village heads in the research sites had shown great interest in the benefits that they could obtain from the activities that were to occur in their areas, and they were keen on having a cut.

Denying this opportunity to the village heads could have meant troubles ahead or, worse, it could have led to the failure of the initiative altogether. In contrast, letting them have a share of the benefits could possibly spare the researchers and the farmer leaders many feuds and tensions with the village officials.

For those reasons, the wisest choice at that time seemed to be to give them a small slice of the benefits. With that in mind, we included the village heads' relatives on the list of sheep recipients, provided that they agreed to abide by the project's policies and rules.

It seemed to us that by doing this, we would save a lot of energy and avoid further conflict between farmer leaders and village heads. This compromise would also make it possible for us to use the village halls for training meetings.

As the activities unfolded, we found out that high mortality rates occurred among the project sheep raised by the village heads' relatives. Why? The answer was simple. They were not farmers and did not have the necessary skills to raise sheep. Moreover, they did not spend enough time on their tasks. Consequently, they failed in raising the sheep and also failed in returning the sheep offspring.

The second problem we encountered was the implementation of inappropriate feeding practices. Farmers were used to feeding natural grass to their animals and gave similar feeds to the new prolific sheep, without adding enough legumes in the rations. As a result, many animals suffered malnutrition.

Malnutrition brought about many miscarriages among the pregnant ewes and high mortality rates among newly born lambs. To solve these problems, the research team introduced urea molasses block as a protein and mineral supplement for the sheep.

The third problem was the lack of grass and forage supply during the dry season. This recurring problem was seasonal and continuously haunted the farmers year after year. At the beginning of the fieldwork, the research team, together with certain farmer leaders, initiated a demonstration plot to cultivate improved grass and legume varieties in the four research sites. However, very few farmers were willing to cultivate the grass and the legumes, arguing that the grass supplies had always been abundant. Why bother cultivating it?

Another difficulty encountered was that farmers did not respond well to the suggestion of silage-making as a way of preserving the grass, particularly during the rainy season, when there is a large supply of grass. They did not like its smell.

Lastly, after the project's end, when we had to reduce the frequency of our visits to farmers due to limited funds, many farmers did not keep their promise to return good-quality sheep offspring to the initiative. This situation badly affected the efforts to revolve the sheep offspring to other potential farmers. The sustainability question then came up. How much longer would the project last?

Despite these shortcomings, this initiative provided a number of learning opportunities. Looking back on the work accomplished, we can draw lessons and identify the most outstanding features that contributed to making it a success story.

First, conducting a comprehensive baseline study on the research sites was very useful in order to understand the situation, the problems and the opportunities related to local natural resource development, as well as the traditional systems and indigenous capacities of local communities. This also allowed for the development of a sound action plan to implement the initiative.

Second, the strong commitment of a funding partner to ensure adequate funding in order to implement the action plan and prepare the communication materials proved to be a key issue. Furthermore, support from local policy-makers, village authorities and farmer leaders also proved extremely important to ensure the successful implementation of the action plan.

The direct involvement of local extension workers and farmer leaders in disseminating the materials, in distributing the sheep and in providing backstopping activities also seems to have greatly contributed to building the necessary trust with local farmers. Moreover, holding monthly meetings proved to be a good way of getting feedback and monitoring activities, while providing further information to farmer groups about sheep raising and marketing their products.

Finally, holding dissemination seminars on the project's success, with the participation of farmer leaders, the academic community and policy-makers, was very important for the project's replication and expansion.

Participatory Research and Water Resource Management: Implementing the Communicative Catchment Approach in Malawi

Meya Kalindekafe

Water resources are so important in sustaining life and livelihoods that their management can be quite complex in light of the numerous and diversified stakeholders involved. However, the lack of a comprehensive and coordinated

management strategy often results in failure to resolve the water-related problems experienced by local communities. Researchers can greatly contribute to changing this situation, provided that they adopt participatory methodologies and do not limit the scope of their research to the biophysical environment at the expense of the social environment.

In Malawi, responsibility for water resources is considerably fragmented. For example, depending upon water use, the departments in charge may be either the Malawi Water Department, the city and town assemblies, the Malawi Fisheries Department or the Ministry of Agriculture and Irrigation. This situation impedes the promotion of integrated and sustainable water resource management schemes. Coupled with inadequate action research, this has led to a failure to resolve most of Malawi's water-related problems, which include the poor management of water resources at local levels; the lack of mainstreaming of gender issues in water resource management; pollution and associated water-borne diseases; catchment area degradation; and the lack of human capacity to assist local communities.

With these problems in mind, a research initiative has been undertaken by the Biology Department of the University of Malawi to investigate the extent of environmental degradation in relation to the use and availability of water in the Lisungwi, Mwanza and Mkulumadzi rivers area. The research initiative also aims at assessing and documenting indigenous knowledge on water resource management, including the coping strategies used locally.

The rivers under study are located in the Mwanza district of southern Malawi, which covers an area of 2239 square kilometres inhabited by 138,015 people. These rivers are important tributaries of the Shire River, an outlet for Lake Malawi and the main source of hydropower in Malawi. The interest in these rivers stems from the fact that they pass through areas with various degrees of environmental degradation, mainly as a result of human activities. Indeed, the rural location of the study area implies that the local population's livelihood relies heavily on the exploitation of natural resources, water being one of them.

In the long term, the study will be useful in developing sustainable and integrated water resource management strategies and fostering economic development in the area. In other words, the results of this study will provide a route map towards designing appropriate integrated water resource management strategies and plans for the area in order to lessen water-related problems and to help achieve local social, economic and environmental goals.

To do this, the study will more specifically try to determine the social and economic influences that should be taken into consideration in policy-making. It will also try to identify the indigenous knowledge that already exists at the local level and that can be used as a stepping stone to develop water resource management plans. It will also pay special attention to gender issues and possible ways of integrating them within the water management schemes. Furthermore, on a more technical level, the quality of water will be assessed in order to determine how safe it is for local communities to use these water supplies as drinking water. If need be, mitigation measures that can be put in place to avoid the degradation of catchment areas will be defined, together with ways of building human capacity at the local level.

Using participatory tools and approaches

In order to enhance participation, an approach called 'communicative catchment' has been used throughout the study. As described by Martin (1991), the communicative catchment approach is an action-based form of research (experts engage in theory-based participatory action with communities) based on systematic thinking (including interrelationship between social and natural environments). It is an approach that allows both experts and local residents to be involved in land-use decisions, as well as in evaluating the long-term effectiveness of their action. As full-fledged participants, communities manage the catchment, while resource managers play a facilitating and coordinating role for community involvement and action.

For a long time, most researchers in natural sciences have concentrated their research on the biophysical environment, while ignoring the social environment. However, experience has shown that the social environment plays a crucial role in the functioning of natural systems. Catchment ecology and management have also provided valuable information for achieving sustainable land-use practices. However, the input and cooperation of local residents is absolutely necessary (Brown and Kalindekafe, 1999). Although various participatory communication or data collection methods have been used over the years to involve local communities and gain their input, in most cases nothing returns to communities once the research project is completed.

In this case, the research team initially conducted a literature review in order to define general guidelines for the study. However, since the root causes of the environmental problems were not known, it was necessary to involve the local community. This was not easy because people have their own priorities when it comes to their livelihood. In order to overcome this difficulty, the study team had to come up with an effective participatory communication approach that could facilitate the relationship with local communities. The communicative catchment approach is being used in this study to ensure maximum involvement and feedback to communities.

The communicative catchment approach, like other like-minded methodologies such as community forestry, is based on concepts similar to those of participatory development communication (PDC), which can be defined as the effective exchange of ideas and information through the active involvement of communities and other stakeholders for improved welfare. In this particular study, the use of the communicative catchment approach is considered a proxy for participatory development communication.

In that context, a number of participatory tools have been used, including questionnaires, focus group discussions, resource mapping and interviews with key informants.

Training

As a first step, enumerators from the study area have been trained in a participatory fashion. The trainees contributed to the improvement of the questionnaires and the development of key questions for the focus group discussions. The training sessions have also allowed for capacity-building in terms of awareness of key issues and the roles that different groups play in natural resource management. Enumerators have helped the researchers by advising them on cultural norms – for example, how to approach women and elders to ensure their cooperation. During the training sessions, the sitting arrangement has been done in a manner that does not place the facilitator as a boss but as a co-participant, which makes the process very interactive. Postgraduate students at times accompany the principal researchers as a way of learning by observing and doing. The knowledge gained from the field experience is discussed further in class and incorporated within their own research projects.

Semi-structured questionnaires

The second step consists of questionnaires being administered for each river. The questions cover all the issues that are crucial to the objectives of the research. This approach requires a lot of time with local people. Enumerators administer the questionnaires under the supervision of the principal researchers. Apart from responding to the questions, people are also allowed to ask questions and make comments on the issues that they deem relevant, even those that are not covered by the questionnaire. This allows for the constant improvement of questionnaires by ensuring that the issues discussed with earlier respondents are not omitted.

Focus group discussions

Focus group discussions, which are open-ended and semi-structured conversations with smaller groups made up of men, women and traditional leaders, have also been conducted. The advantage of focus group discussions is that they encourage the participation and contribution of different interest groups who may not share freely their views and concerns in the presence of other groups' members because of customs and traditional beliefs. One full day is assigned for the focus group discussions for each river.

It is usually difficult to assemble people together in a context where their priority is to solve their immediate needs rather than discuss environmental issues. This is particularly true with women, whose work burden is often overwhelming. The approach adopted in this study consisted of holding discussions as the women continued with their normal activities (in this case, selling goods at a local primary school). The men were easier to assemble in groups.

Key informant interviews

Key informants for this study include retired officers, chiefs, village headmen, field assistants and community development assistants, depending upon availability. Most of these informants are very knowledgeable and cooperative. However, a few of them want their personal agenda to be met rather than that of the community.

Resource maps

As a follow-up to focus group discussions, resource maps are drawn by the same members of the focus group discussions on the following day to examine the different resources used by women, men and other gender groups and the personal and use value that women and men attach to such resources. The purpose of the exercise is twofold. First, it aims to map out the resources that are thought to be associated with dominant socio-cultural categories of 'women' and 'men'. Second, it seeks to map out the spaces used by individual gender groups. This enables the researchers to draw out contradictions between local ideology about gender roles and gender spaces (i.e. what should be) and daily gender practice (i.e. what is). In other words, this exercise reveals local social ideals regarding gender roles and the use of space and resources as it happens in everyday life.

Each gender group is then asked to map out the spaces, places and resources used by women and men using different colours, codes and symbols. Participants are asked to point out and comment upon the key places, features (such as their home or a nearby road), structures and resources that are important to them. They are then asked to identify and draw key places/ spaces that are essential (or peripheral) to their daily activities, as well as the places/spaces which they perceive to be important to men/women and to themselves personally. Participants are not to be interrupted unless they stop drawing, in which case questions are asked to prompt them.

This approach is challenging in that it takes up a lot of time. For this reason, participants are usually given drinks and local food such as *nsima*. Bringing drinks such as orange squash and food such as rice and meat is seen as a special event and attracts people. The meat is bought from the local people and the drinks from local shops. Both researchers and assistants help in the cooking. Although some consider it controversial, the team has discovered that through eating together, people feel that you are part of their group and open up more easily to discussions.

Benefit–analysis charts

The use of benefit–analysis charts, as a point of departure within the focus group discussions, allows for in-depth examination and analysis of who uses and benefits from particular resources. This gives data on who actually benefits from the different resources despite access, control and use. Flip charts are used in the focus group discussions to draw benefit–analysis charts. Researchers explore why women and men use natural resources (i.e. the benefits they receive from particular natural resources) by examining the attributes that women and men ascribe to different resources (nutrition, medicinal use and so on). We also explore who holds traditional knowledge and which resources are commonly sold to local/regional markets and by whom.

Transect walks

Together with the local people, the state of water and of biodiversity is assessed, as well as the general environmental degradation along each riverbank. Local people usually explain the different uses of various species, while the researcher explains the biological and ecological use and the importance of conserving those resources. Both parties learn from each other in the process.

Communication challenges and strategies

In this process, a number of communication challenges may arise. Possible strategies to address the most common challenges are outlined in Table 3.1.

The way forward

The researchers involved in this study have recently joined the Isang Bagsak forum, a capacity-building and networking programme in participatory development communication. Through the various themes posted on the programme's electronic forum, the researchers have been able to exchange ideas and to learn from others on the use of PDC in natural resource management. Ideas that can be applied to the local setting are then communicated to local communities for their comments.

Most people in the study area are poor. They have little formal education and very low levels of basic science, but plenty of traditional ecological knowledge. Therefore, the project will now focus on integrating basic science with sustainable traditional ecological knowledge.

A workshop for the community-based natural resource management (CBNRM) committee members and local extension workers who live and work in the vicinity of the three rivers is planned for the project's end in order to obtain final feedback. The results (including maps produced and photographs taken) will be disseminated to, and finalized at, the workshop. The material will also be used as a basis for future work.

The capacity of local communities is expected to improve through an increase in their knowledge base and the promotion of technological changes in water resource management. As a way of providing feedback to people, maps and photographs acquired during the study will be placed at strategic places throughout the area, such as schools, health centres, churches and

Table 3.1 *Communication challenges and strategies*

Communication challenges	Strategies
Low educational levels, coupled with poverty, result in less appreciation of the need for addressing long-term environmental problems than meeting people's immediate needs.	Before the actual work starts, the researchers spend more time discussing the concept of the study and the benefits of an integrated approach with community members.
Negative attitude towards researchers. In the past, other organizations have interviewed people; but the results have not been communicated back to the communities.	Research team members discussed with the water committees the possibility of identifying places where results could be displayed, but also how they would want the results to be communicated.
Sometimes local activities/events such as funerals affect the planning of activities. When working with a limited budget which does not provide for large contingencies, this can be a problem.	The researcher negotiates with local people an activity date that is not too distant. Researchers have learned to make alternative plans for activities that have to be postponed, such as engaging in further fieldwork or analysing data.
Some areas are difficult to access, so considerable time is spent walking.	Local people advise on viable footpaths to areas. In some cases, researchers start walking a few hours earlier than planned.
Questionnaires take up people's time. Some complain that their time was wasted.	Communication approaches such as focus group discussions are conducted with minimal disruption of daily activities (e.g. at a marketplace).
Rains began late and some scheduled meetings had to be cancelled due to people being engaged in gardening activities.	Interview times are agreed with the village headman.
Communication with the Youth Focus Group, especially with girls, was not possible due to early marriages.	This group is a social category that is no longer consulted and schoolchildren are interviewed, instead, at school.
Some individuals disrupt group meetings.	These individuals are usually disciplined by the local communities.
Few researchers are fluent in the local language.	Researchers make use of local research assistants and teachers.

other common places. It is believed that the visual impressions will constantly remind local people of existing environmental problems and possible solutions. With the sustainable use of natural resources as its objective, the study also aims to build trust in the communities in terms of the project's worth and the relevance of their involvement. As such, those involved in this study are accountable for setting the agenda, disseminating messages, approaching issues, etc. The communicative catchment approach allows for this accountability: once the study is completed, the existing community-based natural resource management committees will periodically monitor the natural resources of the area, especially the 'hotspots', and come up with better ways of managing the environment.

It is hoped that this approach will ensure the integrated management of resources and improve the welfare of local people.

References

Brown, R. and Kalindekafe, M. (1999) 'A landscape ecological approach to sustainability: Application of the communicative catchment approach to Lake Chilwa, Malawi', in FitzGibbon, J. E. (ed) *Advances in Planning and Management of Watersheds and Wetlands in Eastern and Southern Africa*, Weaver Press, Harare

Martin, P. (1991) 'Environmental care in agricultural catchments: Towards the communicative catchment', *Environmental Management*, vol 6, no 15, pp773–783

Communication Across Cultures and Languages in Cambodia

Lun Kimhy and Sours Pinreak

Sharing information or knowledge across cultural and language barriers is never easy. But it can be even more difficult when some of the concepts that one wants to share do not exist in the vocabulary of the other group, or when the latter does not even have a concept for what is being expressed. In such a case, images, pictures and photographs can act as bridges to convey ideas and concepts from one culture to the other. When used by minority groups to translate their worldview into words and ideas, pictures can help to describe the complex web of interrelations that make up their livelihood and improve cross-cultural understanding.

In the province of Ratanakiri in north-eastern Cambodia, the official language is Khmer. On their part, the indigenous people living in the area do not understand that language. Furthermore, they have been living in such

isolation that modern concepts such as laws and land titles do not exist in their vocabulary; hence, they do not have any concept of their meaning.

As part of its Land Rights Extension Programme, the provincial government had a team visit the villages to inform people of their rights and of the country's laws. This was deemed important because indigenous communities were in a desperate situation: outsiders and powerful people had been taking away their traditional land and forests. The communities did not know what to do; but they were worried that their children would be in big trouble. In some cases, newcomers showed them receipts of purchase pretending that these were official documents authorizing the sale of their land. Unaware of the country's laws and of land sale procedures, indigenous communities often believed what they were told and ceded their land. At other times, they were made to thumbprint papers for packets of salt or other goodies, only to realize that they had unknowingly agreed to sell their ancestral land.

Although it had very good intentions, the Land Rights Extension team had a slight problem: its members were all Khmer and did not understand the indigenous languages spoken in the area. Hence, their tools and materials were not suited to what they thought were indigenous people's needs, based on their own understanding of the situation. Most of the time, the indigenous communities did not understand what was being said. Occasionally, when someone was asked to translate the team's material into the local language, the translation was so literal that indigenous people still found it difficult to understand what was being said. Despite these setbacks, the team was enthusiastic and motivated.

Although it was already using participatory approaches in all its development activities, the Land Rights Extension team began looking for alternative mechanisms to increase the participation of communities. At this time, the team's adviser, Sous Pinreak, was invited to join a group which was being trained in participatory development communication (PDC) as part of the Isang Bagsak network. Mr Pinreak became interested in the ideas discussed at the Isang Bagsak meetings and presented them to the Land Rights Extension team, who showed great enthusiasm for this new approach. After discussing it further, team members agreed to try it out.

Deciding on how to go about doing it required further discussion. Team members felt that it would be useful to involve the Community Advocacy Network, which is comprised of community members across the province who have been participating in the formulation of the Forest Law and of the Community Forest Decree.

At that time, the Land Rights Extension team was using a model called the Livelihood Framework as a tool to talk with communities and to analyse their livelihood system. It included four aspects: socio-cultural, economic and institutional elements, as well as natural resources. All these aspects were deemed important for the community to be able to live together (see Figure 3.1). The framework was also found to be very useful in initiating discussions with the Department of Agriculture and for staff members to understand the livelihood of the indigenous communities with which they were working.

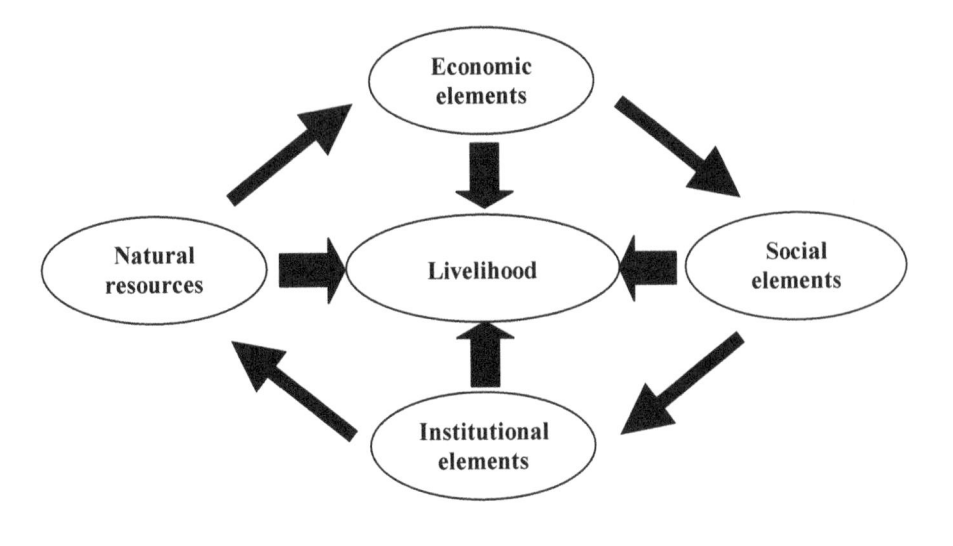

Source: Ratanakiri CBNRM project

Figure 3.1 Livelihood system analysis

The Livelihood Framework was presented to representatives of the network, who also thought it was a good tool to initiate discussions on land and resource tenure issues. However, they hinted that pictures, instead of words, would be a more appropriate tool to illustrate each aspect. Moreover, the selected pictures would have to be representative of indigenous people's reality. Thus, the pictures were drawn according to the advice of network members. Altogether, six sets of pictures were prepared. Each of them represented an aspect of the Livelihood Framework, while the last one was a combined picture representing all aspects. Network members were then asked how these pictures could be used to discuss tenure issues. It was agreed that the first picture would represent natural resources. The process unfolded as follows:

- Communities were first asked what each picture represented and how it related to their lives.
- One by one, the different aspects were presented and people were asked several questions in order to relate the pictures to their lives.
- Next, the picture depicting the five aspects and the interrelation between them was presented.
- As a second step, a picture of the same area, but where all the natural resources had been destroyed, was presented to people. They were asked what was in the picture and what would happen to their livelihood in such a situation.
- Once people had a better grasp of the framework, they were asked about the natural resource issues faced by their community.

- They were then asked if they could identify the reasons for the natural resource issues that they had identified.
- Finally, they were asked if they knew how to solve the problems, emphasizing traditional management systems. Most of the time, people replied that traditional systems could solve only the issues that involved indigenous people, but did not work with outsiders. This opening on cultural differences provided the opportunity to introduce new concepts, such as laws and rights.

Existing similarities between indigenous cultures and the Khmer culture made the process easier. Indeed, both cultures respect age and elders, both want harmony and dislike open conflicts, and in both cases decisions are primarily made by consensus. Conforming to existing social norms is also important in both cultures. Based on these similarities, the members of the Community Advocacy Network made suggestions on how to address the issues with the communities. They also emphasized the importance of using the local language during the sessions.

Once the materials had been developed, the Land Rights Extension team started planning its fieldwork. An idea then came up: why not include indigenous people in the team? Thus, a larger team consisting of indigenous people and government staff was formed. However, some difficulties soon arose: government team members, who were living in towns, had planned their activities without consulting the community members who, for their part, lived in the villages. As a result, the latter were forced to follow the timetable set by the government team, even if they had planned other activities. Other problems also came up with regard to payment and motorcycle use. All these predicaments had to be addressed tactfully so that nobody was blamed and a compromise could be found.

The team then started going to different villages with the aim of sharing ideas and concerns with the indigenous communities, and raising their awareness of new issues such as land rights. The results were very good. In a reflection workshop, community members said that 'facilitation as performed by this team was one of the best'. While in previous activities people would just sit and discuss among themselves during presentations, in this case they listened, asked a lot of questions and greatly contributed to the discussions.

It was interesting to see that community members responded very well to the pictures. Moreover, having the translation done by indigenous members of the team was definitely a strong asset since their translation was not as literal. This was possible because the translator and the trainer were both clear about the objectives of the workshops and about the use of the materials. Thus, they were able to explain the new concepts by primarily using locally applicable examples. This made the team understand the importance of using local language and locally applicable pictures.

Reflections on the use of participatory development communication

Following this experiment, the team had the opportunity to reflect on the process and on the possibility of mainstreaming the use of PDC in land rights extension activities. For one, team members concluded that participatory development communication sometimes involves getting community members to work with government staff. However, the organizers need to be clear about the agendas of the two sides. If they differ, they should be prepared to reduce or resolve these differences so that the two sides can work effectively together.

Furthermore, involving government staff in activities that successfully use PDC improves their familiarity with this approach and builds their confidence. It also increases their motivation to use PDC again.

Finally, using participatory development communication requires that the organizers have resources for transporting participants and paying *per diems* so that community members and government staff can participate efficiently in the process.

Talking with Decision-Makers in North-Eastern Cambodia: Participatory Development Communication as an Evaluation Tool

Lun Kimhy

Most of the time, external experts are responsible for evaluating development initiative, which they do with little or no participation of local communities. Although this kind of evaluation can undoubtedly respond to certain needs expressed by decision-makers or funding partners, it does very little in terms

of capacity-building at the grassroots level. Moreover, the viewpoints and opinions voiced by the communities can be significantly diluted in the process. In that sense, facilitating direct communication between communities and decision-makers as part of an evaluation process can greatly improve its transparency and effectiveness. Assisting communities to discuss and decide on the issues and concerns to be raised, on the best way to express them and on the most appropriate tools to be used can also contribute to building their sense of ownership of the development initiative, while improving local governance.

However, this is easier said than done, and making it a reality can be quite a challenge. In north-eastern Cambodia, the Partnership for Local Governance decided to take up that challenge and apply its newly acquired participatory development communication skills to an experimental evaluation process where indigenous communities communicated directly with government officials. Galvanized by the possibility of being heard at the highest level, participants did not even back up when faced with the difficulty of learning to use high-tech communication tools.

This story tells of indigenous communities who, although they speak very little Khmer (the national language of Cambodia), presented their project evaluation findings to high-level Khmer-speaking provincial government officials. The presentation included recommendations for the officials in charge of a large-scale development initiative, who are more used to telling communities what to do than listening to their concerns.

With a view to responding to the development needs of local communities, the provincial government of Ratanakiri has been implementing a comprehensive, multiyear initiative called the Community Natural Resource Management Project. This initiative aims at assisting highland indigenous comunities in their efforts to acquire greater control over the natural resources they have traditionally used and to improve their swidden farming systems. As such, it supports participatory development activities in community-based natural resource management (CBNRM); improvement of farming systems; local government planning; non-formal education; land titling and conflict resolution; public information; and gender issues. An evaluation is conducted at the end of each year by a team comprised of representatives from the different provincial line departments.

After six months of learning participatory development communication (PDC) through a distance-learning programme, the Partnership for Local Governance (PLG), which plays an advisory role to the project, approached the provincial government to discuss the possibility of a more participatory type of evaluation. Since they had been working together for several years,[1] a relationship based on mutual trust was already in place. Furthermore, these same people regularly socialized outside work on occasions, such as during picnics and dinners. Based on this strong relationship, both sides were used to working with the understanding that the activities undertaken were aimed at building the government's capacities, as well as those of communities. Therefore, the provincial government immediately showed great support for the idea and agreed to allow the communities to evaluate the project.

The various partners involved in the initiative were then asked to identify community members who had experience with workshops and meetings and were more confident compared to other community members. The selected community members were then asked if they were willing to participate in the evaluation process. At an introductory meeting, the director of the Provincial Department of Environment explained that the government had previously performed evaluations, but sensed that some issues raised by the communities had been diluted. This time, the government wanted to hear the voice of the communities directly. Thus, the objective of the evaluation was to empower communities by providing them with the opportunity to express their views directly to decision-makers. Most of the invited community members agreed to participate in the experiment.

After forming the team, government staff, project advisers and the selected community members discussed the indicators that were to be used and the information to be collected. This helped community members to get a better grasp of the project's rationale. They also developed checklists and questions, which they pre-tested. The next step aimed at discussing and deciding on how the information collected would be analysed (see Figure 3.2). At that point, three government staff members helped the community evaluation team, while three members of the Partnership for Local Governance staff ensured that government staff actually coordinated the process without dominating it.

The rationale behind the framework used to analyse the information is as follows: the facilitation of development activities is key to creating a feeling of ownership. Good facilitation, on its part, allows everyone to participate and decisions to be made by all or by a representative sample of the population in a way that is judged to be fair and equitably involves men and women. If these three factors are properly facilitated, the process results in good ownership by the community. Finally, it is believed that good ownership can bring about the changes that the project is trying to achieve.

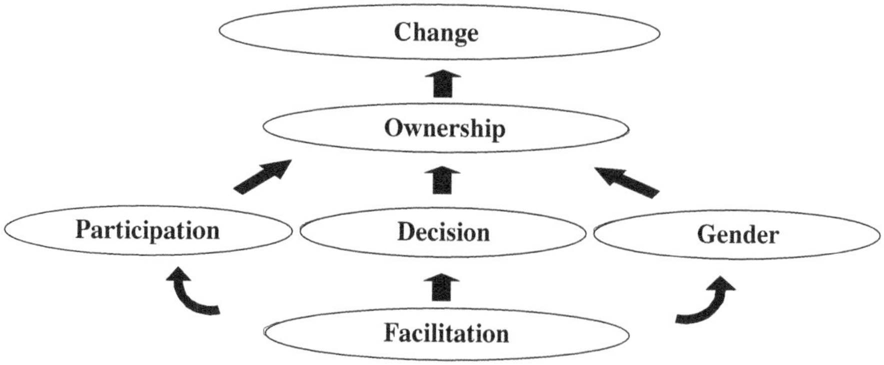

Figure 3.2 Framework for analysing the collected information

After the roles and responsibilities of the evaluation team were agreed upon, information was collected from 15 villages. The data was organized under the following headings: facilitation, decision-making, participation, gender, ownership and change. The team then summarized the data under each heading. The next step consisted in asking the team to reflect on what the information told them about the different headings.

The main findings indicated that, most of the time, government staff, officials and workers use the Khmer language, are impatient and don't listen to the opinion of communities. As a result, activities are mainly determined by government. The evaluation team requested that the project's goal and objectives be explained succinctly, using the local language as much as possible. Furthermore, government staff should consult and discuss clearly with community members before making any decision. They should also agree on a schedule of activities well ahead of time.

The communities then had to present these results to the government. At that point, it became important to consider the tools and methods that could possibly be used to convey the information in a manner that would be acceptable to the participants and within the capacity of the community. Based on a methodology typical of communication strategy design, the evaluation team was asked to consider who the audience was and what attitude it typically holds towards the communities. The issue of who should do the presentation was then addressed, as well as its design. Where did they think they could have problems? This showed that even though communities were able to produce good results, it was equally important to present them properly and to define the basic elements of any communication strategy.

It was decided that the indigenous members of the team would present the findings. Each team member was assigned a different part. Having decided that they would take turns in presenting their findings, people wanted to use simple, short and sharp messages with a clear sequence and flow. They also wanted pictures to be part of the presentation. Having recently bought a new liquid crystal display (LCD) projector and having experimented with Power-Point presentations, the Partnership for Local Governance representatives hinted that a PowerPoint presentation could be appropriate for this situation. The presentation was then developed jointly, and community members immediately tested it. The testing included questions by the advisers and by other team members.

During the workshop, the team made an effective and clear presentation. Government officials were very impressed since they never expected community members to be able to use this technology. Community members were able to answer all the questions raised by the government officials, who were very interested in finding out how the evaluation had been done. The two parties understood and listened to each other. This showed that the normal sender–receiver (government–community) relationship can be transformed when the receivers are in charge of the communication process.

The PowerPoint presentation was very new to Ratanakiri and had the audiences glued to their seats. As a result, everyone paid close attention to what was being presented. One of the lessons learned from this initiative is that, sometimes, new technology can be used very effectively to get messages across and to help people forget who the presenter is. But, more importantly, their participation throughout the evaluation process allowed community members to become more confident, capable of defending their findings and of clarifying what they had done. The practice sessions and going through the analysis step by step was very important in providing this confidence.

In general, the officials accepted the findings, with some changes. After the workshop, the project's information and education component began producing audio cassettes in the local language; translation into the local language became more frequent during discussions, and communities were allowed to discuss among themselves in their own language.

It is not completely sure that these changes were brought about only by the evaluation findings since a number of other factors were also influencing the provincial government at the time. Various organizations were, indeed, insisting on the use of local languages, on the importance of being patient and on the need to listen to community members.

To assist indigenous communities communicate with high-level provincial government officials requires time, resources, energy, patience and effort. Management can facilitate this process by developing a good working relationship with government partners, holding regular meetings with them and using project documents to get them interested in trying out better ways of implementing development initiatives and building a stronger relationship with community members. Management must also support community members as they learn new concepts and experiment with new technology.

Note

1   Advisers had worked with the government officials in the Cambodian Relief and Reconstruction Programme that had preceded PLG, as well as with two previous phases of an International Development Research Centre (IDRC) project.

From Rio to the Sahel: Combatting Desertification

Ahmadou Sankaré and Yacouba Konaté

It is estimated that 900 million people around the world are affected by the phenomenon of desertification, which is bringing with it drastic changes to the climate and to human activity. In 1994, in follow-up to the Earth Summit that was held in Rio de Janeiro two years earlier, an international Convention to Combat Desertification (the CCD) was signed. That event marked an important step in the worldwide effort against desertification. Of course, agreements negotiated in grand international forums do not always

have the hoped-for impact at the local level. Nevertheless, this convention represented an innovation in that it involved local people from the outset as the driving force in its implementation. It was in this context that the Permanent Interstate Committee on Drought Control in the Sahel (CILSS) set out a few years later to develop and try out a participatory communication approach. The idea underlying this approach is that those who see their food security threatened and the biodiversity of their lands impoverished are best placed to design and implement initiatives to protect their environment.

The idea emerged at a time when most countries of the Sahel were working to design their own national action programmes for implementing the CCD. Recognizing the importance of a participatory approach and of forging partnerships with all stakeholders and, most importantly, with the local people, the member countries of the CILSS attempted to find new ways of doing things. An analysis of the strategies used during the past showed clearly that their top-down approach severely limited their impact. By focusing on disseminating information and exhortations, those strategies missed the key point, which is to involve the people themselves.

The experiment with participatory communication as a methodology was conducted in two member countries of the CILSS: Burkina Faso and Chad. Seven sites were selected for the trial, four in Burkina and three in Chad. The experiments were designed to attack problems of great diversity. Yet, all were directly related to the phenomenon of desertification and its consequences on people's daily lives. This chapter discusses four of these experiments: two from Chad and two from Burkina Faso.

Doum-Doum: Halting the desert's advance

Located some 200km from N'Djamena, the capital of Chad, the sub-prefecture of Doum-Doum covers an area of 2647 square kilometres. Its population is estimated at 1000 inhabitants. The locality faces a problem with the sanding-up of its polders and wadis, the main zones suitable for farming. In fact, like most of Chad, Doum-Doum is a very arid zone, almost a desert. The polders (dyked lands) are the beds of dried-up lakes where the soil is highly fertile, containing a great deal of organic matter. The depressions between the dunes, known as wadis, have a silt-clay soil and a shallow water table, which makes them highly productive as well. The encroachment of sand into these lands thus constitutes a real threat to local food security.

It was in this context that the Doum-Doum rural development project undertook its experiment in cooperation with the CILSS participatory communication project.

The initiative began with a series of information sessions and discussions in order to identify the problem and to sound out local ideas for solutions.

To do this, the facilitators first organized information and discussion meetings with the key opinion leaders. They then visited the polders and the wadis in order to assess the scale of the problem. Next, they held chat sessions with local people to analyse the problem jointly and to propose solutions.

At the end of these discussions, people took upon themselves the task of reforesting their environment by growing 10,000 tree seedlings for each zone, to be transplanted when the rains came. Farmers were careful to choose species that were well suited to the zone and resistant to drought.

These chat sessions served to 'loosen the tongues' of local people and encourage them to express their needs – for example, the need for wide-diameter wells for themselves, for their livestock and for the plants that they would shortly have to start watering. Thanks to these sessions, the villagers also saw the advantage of organizing themselves in groups. Today there are 20 women's groups, whereas there were only 13 at the outset.

One activity involved visits to other zones and sites as a way of making local people familiar with solutions adopted by communities facing similar problems. Two delegates from each zone travelled to the sub-prefecture of Ngouri, where they were impressed with the successful use of hedges or 'live fences' and decided to replicate that experiment at home.

People also recognized the need to strengthen their capacities. The facilitators therefore arranged training sessions on sowing and watering techniques. At each session, the facilitators used flip charts to illustrate their points. This strategy served to trigger in people's minds an awareness of the role that each could play in halting desertification.

Moreover, during the sessions, the older people of the village were able to recall the environment as it was during the past when, although it was threatened, it was not so severely degraded. This provided a lesson for the younger generations who are gradually losing their farmland to the advancing sands.

The training session featured films on the environment, organized for each zone under the eye of the facilitators. Here, again, people were prompt to react: one 60-year-old man from Kouloudia exclaimed, after watching one of the films, that 'Sanding is not inevitable; we know that there can be solutions even here.' The solutions that these farmers are now investigating involve the planting, tending and maintenance of plants around their fields, whether in the wadis or in the polders.

Participatory communication thus helped the communities to identify their problems and analyse them, while bringing unspoken needs into the open and facilitating home-grown solutions.

The principal outcomes at Doum-Doum

Although the extreme heat and lack of water were hard on the seedlings, of the 30,000 planted in the nurseries, 10,341 survived and were transplanted to fix the dunes. The area ended up with 8km of plantings, four to five rows deep.

The experiment with participatory development communication (PDC) also highlighted the damage caused by cutting trees for firewood and the importance of using substitute fuel sources. Dialogue on these questions addressed the use of improved wood stoves and the burning of cornstalks as a way of saving wood.

The initiative also made farmers aware of their capacity to analyse problems and to find local solutions. This sparked in many of them the determination to take their fate into their own hands and to plan for the future. They also came to appreciate more thoroughly the value of natural resources and to draw up local regulations to combat desertification. In fact, the communities showed a real desire to be active rather than passive players.

Bol: The silent revolution

In the area around Lake Chad, there is no clear relation of ownership between the land and those who work it. The area is heavily Muslim and the land belongs theoretically to God. In fact, it belongs to the first village or clan that settled it. Villagers may have user rights to lands located on the dunes, in the polders or in the wadis. There are apparently no sales of land, which are, in fact, prohibited.

Outsiders have the right to cultivate vacant lands; but they have no permanent user rights. They can, however, acquire the land if they live long enough in the locality or if they marry a girl from the village. They then become full members of the community and this status gives them definitive rights to the land.

The situation is quite different for women, even those born in the region. Although they are actively involved in all stages of farm work alongside their husbands, they are seldom awarded title over a plot. Apart from working on their husband's land, their main input is confined to harvesting the crop, caring for the animals and selling livestock by-products. It is not for any shortage of land that this is so, and some of the polders developed by the Société de Développement du Lac, in particular those at Berin and Guini, are still underused and even unoccupied. Women could make use of these lands and contribute to family income and to the community's food security, while enhancing their own economic independence. Yet, tradition insists that women have no right to the land and this is holding back the development of the villages around these polders.

It was this very complex problem that the participatory communication initiative attempted to resolve. The results that the facilitators achieved, by sparking real dialogue on this issue in the villages, are nothing short of astounding. Some people even speak of a silent revolution.

Women as landowners

'Now we talk openly about things, whereas before we tended to keep quiet. The women's group that this project has helped to set up can only make us even more open.' Those were the words of a woman in Sawa, one of 30 small villages clustered around the polder, which is now being improved. The women of this village are making plans for the future: 'With the income from our plots we are going to buy a mill, a seed drill and some threshers.' The women of Bol now have full rights to the land and they are determined to become increasingly independent and take care of themselves. Were it not for the delay in building the irrigation ditches, they would already have their plots. They have been assured, however, that the plots will be distributed equitably once the works are finished.

The communication strategy

The communication strategy that brought the village to this point unfolded in several stages. First, the facilitators secured the support of the traditional authorities, particularly the Sultan of Bol, who is greatly revered by local people. He agreed to go up and down through the villages in an effort to get the men to change their minds about making room for women in their proceedings.

Enlisting this traditional personality was helpful in obtaining husbands' consent for their wives to take part in the initiative. In many parts of Africa, a woman cannot make any public moves without the consent of her husband. At a public meeting, women may accept a proposal; but until their husbands have given approval, they will not do anything. The Sultan's involvement helped to overcome the husbands' initial reluctance, and they were readier to accept their wives' participation.

The facilitators next held a series of meetings with the husbands, and then with the women. After a few chat sessions the husbands finally agreed that turning over the plots to their wives would be a plus for the family budget.

Yet, despite this consent from their husbands, women's right to speak out was not yet assured. This was something that developed over a series of stages. As always, at the initial discussions it was only the head of the women's association who spoke: what she said had an 'official' ring to it and was not particularly interesting. In such a setting, the spokeswoman says only what has been agreed, which is what her audience wants to hear. The same was true for the next three or four sessions.

Among the women of Bol, the floor at a meeting is awarded by seniority. The women's president is often the oldest woman of the village or someone who is close to the chief. The ones who follow her in speaking will be the older women. Their interventions cannot last very long, two or three minutes at most, filling a total time of 20 to 45 minutes.

The transition between this and the second stage is rather delicate. It depends upon several factors: the degree of trust that the facilitators have won, their tactfulness and their familiarity with the setting. The facilitator must be aware of local traditions and he must speak in proverbs and allegorical turns of phrase in order to put the crowd at ease and encourage them to speak.

It is during the third phase that the real give-and-take occurs, and this can happen without warning. It is noticeable when people begin to draw out their statements (from two or three minutes at the beginning to perhaps ten minutes) and when women no longer wait to be given the floor. A greater variety of topics are then addressed; people will speak about anything that comes into their minds. It is at this point that the women begin to discuss their problems and put forward their thoughts. At Sawa, it took more than an hour to reach this point. All those who spoke at the beginning paid reverence to the Sultan, betraying his considerable influence, before they got to the problem they wanted to raise. Then came a point when three women jumped up at the same time to speak, and the meeting had to be called to order so that everyone could have a turn.

One of the sessions ended with a spontaneous popular dance. This was something the husbands had not seen for a long time. A *marabout*, or holy man, confided to one of the facilitators: 'Your coming has liberated our women; we've never seen them that way before!'

Some valuable lessons

The village of Bol has decided to address a very sensitive problem: the marginalization of women in the distribution and farming of improved lands.

Our experience showed that when communication tools are properly used the community becomes a frontline player. In these cases, the participation of traditional authorities such as the Sultan, the district heads, the women leaders, the *marabouts* and the village chiefs was crucial. It was they who facilitated the meetings with the women and took part in them initially.

This approach did not involve any mass communication tools such as radio or television. The emphasis was on chat sessions, supported sometimes with photographs, videos or tape recordings dealing with experience in neighbouring communities.

One of the most effective communication tools turned out to be photographs from the discussion sessions. They were posted in the village square where everyone, and especially the women, could see them and could point to themselves taking part in debates about their own very real concerns.

The tape recorder was always at hand for recording the discussions, and listening to the cassettes afterwards sparked very positive reactions. The showing of films, followed by debate, encouraged people to speak. After viewing a video on how women in another region had secured rights to land, one woman in Sawa declared to the audience: 'We want to be like our sisters in the village of Matafo, who organized themselves and won access to land. Now they can use those harvests to meet family needs without turning to their husbands.'

Beli: Managing common pasturelands

The Beli transfrontier project is attempting to set up a suitable framework for the shared use and management of pasturelands. The idea was to facilitate communication as a means of creating this framework. In the end, the project led to the establishment of decentralized committees in the villages.

A fragile ecosystem

The district of Tin-Akoff lies along the border with Mali to the north and Niger to the east and has some of the best pasturelands in the province of Oudelan. The Beli River, which flows through the district from west to east, is the focal point for herders.

The river supports shrubby thickets all along its banks and affords rich pasture for the herds. Once the haunt of wild animals and visited only by the boldest herders, the land today draws herds from other districts of Oudelan and from other provinces, as well as from villages in the neighbouring countries.

The sparse, shrubby vegetation of this chronically rain-deficient region is at risk of rapid degradation from overexploitation. The river, which once ran deep, dries to a braiding of trickles by January because of the invasion of sand and the great numbers of livestock. Its banks are increasingly being taken over by fields and gardens. The lack of defined livestock paths to the water means that the whole ecosystem is heavily trampled.

Although hunting was once a favourite pursuit, the district is now losing its wildlife. One of the explanations proffered pointed to the widespread poaching that has decimated the population of wild animals or driven them to take refuge in neighbouring countries. Yet, there are still some flocks of gazelles to be seen and the occasional hyena or jackal, as well as many species of birds.

There are two social groups that pasture their animals in Tin-Akoff: the Kal-Tamachek and the Fulbe (or Peul). They have long shared the area and they both speak the same language, Tamachek. Herding is the main source of livelihood, but most people also do some farming. The area around the village is therefore divided into areas for pasture and for crops.

The fields were traditionally located upland among the dunes; but with recurrent droughts, they are now to be found increasingly along the banks of the river. As for the pasture areas, they shift from the dunes to the river, depending upon the season.

The communication strategy

The communication strategy has three main objectives: to share knowledge about natural resources, to introduce localized frameworks for cooperation and to make those frameworks more effective. The main partner responsible for carrying out the transboundary Beli project was the Waldé Ejef Association. It selected as facilitators men who were familiar with the locale and women who were socially respected (one of them came under the authority of the traditional chieftainship). The only drawback was that they were not based in the villages but came from the departmental capital.

To achieve the first objective, the facilitators and the villagers used chat sessions, posters and videos as their tools. Two films provided backup for the discussions. One dealt with the Oursi pond and the other with the management of hunting areas in eastern Burkina. Accompanied by local foresters (the facilitators are not natural resource experts), they engaged people in discussion about the status of their resources, the pressures acting on them and the best ways of preserving them.

The facilitators succeeded in getting the villagers to talk about what was happening to their natural resources, about their relations with migrant people and about the unbridled exploitation of vegetation and wildlife. In all the villages, people complained that wildlife was disappearing because of illegal hunting and they were disgusted with poachers who would arrogantly brandish a hunting permit issued by the authorities. They were unanimous on the need for committees to control hunting.

Taking the bull by the horns

The experiment at Beli did not produce any results of a physical kind. Yet, the facilitators' efforts encouraged a movement to social organization: in the eight villages there are now associations of farmers, herders, women, youth, and wildlife and fishery management units. Organizations of this kind can help to sponsor development activities. In some villages, steps have been taken to create tracks for livestock to reach the river. During the meeting with government officials and the project coordinators, people let the authorities know that they wanted to be involved in environmental management. Initiatives are now under way to obtain permits for wildlife and environmental management. If those efforts succeed, they could well lead to income-producing activities. In short, local people and government extension services are now working closely together.

Not everything has run smoothly ...

Despite these encouraging results, the Beli initiative ran up against obstacles that hobbled its activities. In addition, the distance of the facilitators from the field (at the start of the project only one facilitator resided in Tin-Akoff, while the others stayed in Gorom-Gorom, about 70km away) severely constrained contact with local people. Moreover, there were not enough facilitators (only three for eight widely scattered villages); even worse, the first group of trained facilitators had to be replaced, in time, by others who had received no instruction in the approach before they were dispatched to the field.

The lack of commitment on the part of the authorities also led to some problems. Although they were fairly familiar with the project and its objectives, they were still confused about the difference between the Beli project, Waldé Ejef and the participatory communication initiative. This confusion was not of great consequence because all players were working towards the same end. Government officials, however, complained of the lack of information and the shortage of financial and material support, which sharply curtailed their participation. In fact, when the authorities are completely empty handed (the provincial livestock office has no electricity or telephone, let alone vehicles and fuel), their field presence will be very limited.

Ouarkoye: Fighting bushfires

Every year fires consume 30 per cent of Burkina's territory, with devastating impacts on vegetation, soils, animals and people. The district of Ouarkoye, in the region of Boucle du Mouhoun, is no stranger to this scourge.

The most severe consequences of bushfires include the following:

- the decline in forested areas resulting from bushfires and from the expansion of the farming frontier;
- the inability of vegetation to regenerate itself because of persistent wild-fires;
- accelerated erosion along watercourses;
- reduced harvest yields;
- loss of soil fertility;
- the great frequency of fires (61 per cent of the surface area is burned every year); and
- shortage of pasturelands.

The communities in the district of Ouarkoye, like most of those in the western part of the country, are headless societies in which the chieftainship has little hold on the population. Decision-making power in all matters, including natural resources, lies in the hands of the land chief, who parcels out the land, particularly through indigenous families. Traditional customs remain strong and exert a firm hold on social life.

The traditional channels of communication operate through the markets that are held once a week or every three days, or through places of worship, events such as marriages and baptisms, and the many snack bars in the area.

The people of Ouarkoye, faced with the degradation of their natural resources, could never agree on how to plan and implement a proper fire management programme. Consequently, they seemed to have little interest in protecting their resources.

The communication strategy

The first step was to discuss with the villagers the causes and consequences of bushfires. After making initial contact with the customary leaders in the district, the facilitators turned to a theatre troupe to dramatize aspects of the issue by means of a play. This production toured the villages, sparking debate on the issue of fires and encouraging people to speak about it.

At the same time, the facilitators and the fire management committee heads met with different social groups such as women, village elders, leaders, young people, teachers and hunters to provide them with information and gather their opinions.

The local radio station in the province also helped to convey information and to feed discussion by broadcasting a series of programmes on the fire problem. These were transmitted in the local Dioula language and they struck a real chord with the villagers, some of whom went to the station for further information.

Gradually, the leaders of the village fire management committees, and then the general public, learned about the techniques of fire management.

Working with the traditional chiefs

The traditional chiefs played a key role in the project's success. Their support and involvement quickly won over all the stakeholders. The same could be said of the use of theatre – indeed, the village troupe that was forged during the first performances went on to carry the show to many other villages.

Results and constraints

At the end of the project, each of the 28 villages and farming settlements had established its own fire management committee. Subsequently, two training centres were set up and are now operational. They have trained 362 instructors, who in turn have provided training to the villagers.

Among the initiative's most important outcomes were the creation of a local theatre troupe (called Sininyasigi) and four community activity centres, as well as the production of eight programmes dealing with fire management, which were broadcast over the local radio station Fréquence Espoir.

According to one facilitator: 'With forum theatre and the video shows, farmers were able to see what it's like to live where there are no fires. They could compare that situation with their own burned lands.' Members of the forest management committee at Miena (a village surrounded by high grass that was thoroughly dry at this time, in March 2002, something that was unimaginable before the PDC project) explained: 'The soil has been enriched because where there's been fire, and where there hasn't, it's not the same thing at all!'

Apart from these concrete outcomes in terms of communication, the experiment also produced the following results.

A significant reduction in bushfires

There was a sharp decline in bushfires in most of the villages where the project was undertaken, thanks to action by the villagers themselves who built firebreaks several kilometres long to protect the sacred forests, the villages and the fields. As an indication of the improved environment, a number of wildlife species that had disappeared have now returned to the area. In August 2001, an elephant moved into a forest protected by the villagers, and this was quite a novelty.

A qualitative change in the relations between stakeholders

The project brought about a marked shift in the attitudes of the facilitators, the forestry officials and the local populace alike. 'In the past', admitted one forest officer, 'we acted like policemen over the peasants. We would call the farmers together and tell them things, and that was all. It was a completely unilateral approach. Today, it's the villagers who explain what has to be done or not done, thanks to the stage play.'

'Before, if a farmer saw a fire he would keep mum about it', said one of the facilitators. 'Today, the villagers are quick to sound the alarm and go put it out.' The facilitators also insist that, before the project, it was hard to find the people who set fires; but that is no longer the case today.

Farmers care for their land and help the forestry officers

Today, farmers feel responsible for their land and they look after it diligently. The facilitators found that farmers are fully committed to managing fires on their land. They settle minor legal disputes by themselves, with the help of the forestry officers. They set and enforce fines, a portion of which goes to the village. The traditional chiefs establish special areas for ceremonial fires, which are now kept under strict control. The village of Kebaba has even created a 2000ha forest reserve, and most of the other villages are hoping to emulate that initiative.

Renewed dialogue and trust between the forestry service and local people

Before the forest management committees were established, forestry officers were somewhat lax in performing their surveillance duties. This has now changed. 'No longer does a forestry officer simply draw up a programme alone in his office and then put it to a sparsely attended village meeting. Now, it's the fire management committees that organize the meetings. The people are more comfortable in talking about things and they take an active part in the debates', says one facilitator.

Stronger partnerships with the technical experts

Partnership with the main technical services (farming and livestock, the environment) has been strengthened and extension workers are now given training in the local language.

Communication tools at the service of local people

Throughout this PDC experiment, we used simple, inexpensive and locally adapted communication tools. We did not resort to mass media, such as radio or television, but relied on the local tools at hand. In all cases, the facilitators made local people the focus of their strategies, using home-grown modes and channels of communication. The emphasis was on maintaining contact with the local traditional authorities and on conducting chat sessions, sometimes supported with photographs, videos and tape recordings from experiments in neighbouring communities. Moreover, the experience showed that when tools are carefully selected in light of objectives and are placed in the hands of local people, it is they who become the principal players in development.

The first phase of this action research project wound up in May 2002. The approach taken and the successes achieved were analysed during a regional roundtable that brought together stakeholders from member countries of the CILSS with an interest in this approach. Convinced of its utility, all of them expressed their desire and their intention to extend this experiment to other member countries of the CILSS.

Engaging the Most Disadvantaged Groups in Local Development: A Case from Viet Nam

Le Van An

Policy changes aimed at protecting the environment can have a major impact on local communities when it implies radical changes to their way of life. This was the case in central Viet Nam, where the government established new regulations in order to better protect the forest. For the ethnic minorities whose livelihood was based on swidden farming, this meant that their ancestral knowledge was no longer useful in the new context. This could have signified further marginalization and impoverishment of the local population. However, the use of participatory methodologies and, later on, of participatory development communication (PDC) made it possible to engage local farmers

in a process aimed at discovering and trying out new ways of making a living, while safeguarding the natural resource base. Special attention was paid to those sectors of the population who usually do not benefit from development initiatives, with surprising results.

In Hong Ha, a commune located in the province of Thua Thien Hue in central Viet Nam, the livelihood of the upland minority people used to be largely based on slash-and-burn cultivation and on the natural resources from the forest. However, during recent years, a new government policy has been put in place to encourage these farmers to change their practices from swidden to sedentary farming in order to improve their access to social services and ensure the protection of the forest. As a result, they have had to change their agricultural systems. This is not easy for them as their indigenous knowledge and skills are no longer relevant in the new production system.

The main problems faced by the upland people in central Viet Nam are poverty and the degradation of natural resources. The lack of food causes serious hardships to families since many of them do not have enough rice to eat for three to five months every year. Cassava and upland rice used to be their traditional food crops; but the population growth, combined with the limits of swidden farming, has resulted in a sharp decrease in food production. Although wetland rice is being replaced with upland rice, the paddy field is small and there is no water supply for the cultivation of rice. As a result, during the food shortage season, poor farmers have to go to the forest to collect non-timber products and sell them in order to get enough income to buy their daily food.

Moreover, the area's natural forest was seriously damaged during the war due to fire and the use of the chemical defoliant known as Agent Orange. After the war, when people started to rebuild their lives, trees were cut down to be used as construction materials and to be sold for income. As a result, the quantity and quality of forested areas have greatly decreased over the last decades. Since the early 1990s, the government has been making great efforts to invest in forest management and has undertaken a number of replanting forest programmes. In order to support these forest and land management efforts, a forestry law was issued in 1991, followed by a land law in 1993. Thereafter, the government has made a number of decisions with the aim of increasing the rights of farmers with regard to land and forests. However, the necessary conditions to apply these new policies are not yet in place in the upland areas due to the poor management system of local organizations and the lack of understanding on the part of the local population.

As a matter of fact, most of the land in Hong Ha falls under a watershed management scheme. Local people only have access to about 1 per cent of the total land area for agricultural production. Given the population growth, the changes introduced in agricultural production techniques and their limited access to resources, one truly wonders how the upland people can possibly improve their livelihood and their resource management practices with a view to sustainable development.

Hence, given the existing needs in upland central Viet Nam, since 1998 the University of Hue has developed a research initiative entitled

Community-Based Upland Natural Resource Management. This initiative has been implemented by the research team of Hue University of Agriculture and Forestry. It aims to:

- improve the livelihood of the upland poor;
- advance human resource capacity; and
- support policies that perform for the poor in the upland areas.

From the onset, the methodology of participatory action research has been used to work with farmers in order to identify problems, develop potential solutions and monitor the changes and the impact of the initiative on their livelihood, as well as on natural resources.

In 2002, participatory development communication was introduced to this initiative through a learning process that involved other teams in Cambodia and Uganda. The community-based natural resource management (CBNRM) initiative in Hong Ha was chosen as the study site in Viet Nam, where this new approach was to be experimented with. Henceforth, its researchers were trained to use its methodology and tools.

The research site

The Hong Ha commune belongs to the A Luoi district, which is located in the province of Thua Thien Hue in central Viet Nam. There are 21 communes in the district, of which Hong Ha is one of the poorest.

The commune has 230 households and approximately 1200 inhabitants, who pertain to different ethnic groups. The K'tu people are the most numerous, with 47 per cent of the total population. The Pa Co (including the Pa Hy) account for about 28 per cent of the population and the Ta Oi for 16 per cent, while 7 per cent are Kinh or lowland Vietnamese. The official land area that belongs to the commune covers 14,100ha, of which only 180ha are agricultural lands. Forestry lands cover about 11,000ha; but not all of it is considered 'good' forest. There are also bare hills, which cover about 2700ha. Most of the forestry land is under the management of the State Forestry Enterprise and the Watershed Protection Board. Previously, local people practised shifting cultivation in these forest lands; but they have started to turn to sedentary farming since swidden cultivation is no longer allowed by the local and national governments.

Due to these forced changes in their farming system and their limited access to assets and natural resources, production is low and the opportunities for income generation are very limited. Traditionally, the livelihood of these upland people was based on forest resources such as non-timber forest products for food and cash income. However, this is now a minor source of income to respond to their needs as the forests are becoming more and more degraded. Furthermore, the changes in policies and in the management system for these areas have also restricted the access to these assets.

Process, results and impact

The CBNRM initiative aims to develop a research methodology based on participatory approaches. Capacity-building of local people is also a major objective, with a view to increasing their participation and that of other stakeholders throughout the process. Participatory action research is a method aimed at developing a learning process for both farmers and researchers. Besides this new research methodology, the research team was trained in PDC through face-to-face and internet-based learning involving researchers from Viet Nam, Cambodia and Uganda. Nine learning themes were introduced, from approaching a community to developing and implementing a plan, monitoring the unfolding of activities, and evaluating and disseminating the results. In each team, researchers organized as a group to discuss each topic and to post their knowledge on a website. Thus, they learned from the experience of other teams and shared their own experiences in implementing participatory development communication. Indeed, for each theme that was addressed, researchers not only learned through theory, but also applied the knowledge to their research work in the field. Half way through the process, all members of the three teams gathered together in Hue to share their experiences. The final workshop was held in Uganda, where the whole process was evaluated.

In the context of the CBNRM initiative in Hue, the integration and application of PDC has increased the participation of women and the poor in order to improve food production and income generation. It also contributed to building the capacity of local people, especially that of women and disadvantaged people. Furthermore, it led to increased access to resources for local people, especially to land and forest resources. Finally, it put in place a learning and dissemination process to share the results of the initiative with other communities.

Increasing people's participation, especially among women and the poor

It is not easy to achieve 'real' participation by villagers and stakeholders in the whole process in a community-based approach. For many years, in Viet Nam and in other developing countries, technologies or solutions for agriculture and rural development were introduced by outsiders, such as scientists or development workers. The level of farmers' participation was very low. Recently, given that the 'top-down' approach has not yielded real results in terms of rural development and is increasingly being rejected by farmers, the participatory approach has been gaining ground. 'Real' participation can be induced by using different approaches and methods to meet the practical needs of farmers, as well as to enhance their confidence and their own human capital. Furthermore, the use of PDC in natural resource management helps researchers to better understand the situation of the people with whom they work.

Who participates in research processes?

Not all farmers have the same interest and conditions with regard to participating in an initiative. Usually, the advantaged people or farmers who are not poor and have better social capital in the community tend to participate more. Women and the poor, as 'marginal people', are rarely involved in these activities.

There are many ways to approach a community. Participatory development communication aims first and foremost at improving the understanding of researchers. Making a list of the farmers or having them classified according to their wealth levels by community leaders and by different groups of people can help in understanding the community. Each group of people has its own problems and interests. Communities should be approached in an appropriate way, compatible with the existing cultural, social and economic conditions. The more we understand a community, the better the participation will be.

Why do farmers participate?

There are various reasons why farmers participate in development initiatives. In many cases, farmers want to improve their production or income. However, in some instances, they simply want to be involved because development projects can bring them direct benefits.

At the beginning of a research initiative, it is very important to undertake a diagnosis of the situation and to define the research issues. The meetings with villagers should be organized separately for different groups, such as women, men, the poor and leaders. Since the results from these meetings will all be different, researchers and villagers will be able to better understand the situation and to develop the following steps.

How can participation be increased?

Participatory development communication aims to improve the quality of participation. Throughout their training, the researchers had the opportunity to acquire the necessary knowledge and skills to improve their capacity to encourage farmers' participation. They achieved this by developing communication strategies and using appropriate tools, such as video, cameras and leaflets. Learning by doing in the field is also useful in facilitating the involvement of farmers.

The type of questions and the way in which the dialogue is established are also important when talking with farmers. In-depth questions are used to help understand the situation better and to figure out what the farmers' ambitions are. Usually, open-ended questions are used to gather general information. If more information is needed, in-depth questions and discussions come in handy.

Many poor farmers and women benefited from the use of PDC in our study. Numerous farmers said that they had never participated in community development before as they thought that it was not an opportunity for them. Since PDC has been introduced, local people have had greater opportunities to participate in development initiatives.

Improving the local livelihood

Meetings were conducted to identify the needs and problems of local people, as well as the possible solutions to overcome these problems. Since people have different farming activities, meetings were organized with different groups: leaders, women, the poor and men. Each group had different opinions regarding priority activities that should be undertaken.

Therefore, based on the capacity and interests of farmers, different working groups were formed: a rice production group; a pig production group; a fish raising group; a home garden improving group; a cassava production group; and a forestry production group. The research team worked closely with these groups to help them understand that the research initiative was not a development agency, but that it could try to help them by improving the quality of participation.

Each group consisted of a number of farmers. In general, there were no more than 15 people per group to start with. Farmers discussed their problems and the possible solutions that they envisioned might solve them. With the facilitation of researchers, discussions also took place in groups so that an overall diagnosis of the situation could be suggested.

The next step consisted of developing possible solutions, based on the causes of the problems. For example, in the case of rice production, farmers decided to test three new varieties of rice and to try out the application of a fertilizer, as well as transplanting and direct-sowing methods.

In each experiment, about three to five farmers agreed to test the selected options. The group itself made the decision as to who would get to apply the different tests, based on their interests and on the group discussions. The other group members were invited to participate in the evaluation and learning meetings at least three times during the growing season: once at the beginning during the experimental design stage, the second time during the growth period and the last time at harvest time. At each meeting, farmers developed their own criteria to monitor and evaluate the results of the new technologies and made the decisions concerning which varieties were to be sown, which fertilizer was to be applied and which cultivation techniques were the most appropriate. The results from on-farm monitoring and evaluation were shared with other group members and with non-participants. Thus, the learning process also encouraged farmers to disseminate the results to other farmers in their community.

Similar steps were developed with the farmers interested in pig production and fish raising or in other types of activities. In the case of pig production, three experiments took place. The first experiment aimed at raising Mong

Cai as mother pigs or as sows, while the second one focused on raising cross-breeds and fattening pigs. Finally, an experiment was conducted on ensiling cassava root and leaves for feeding pigs.

However, some activities were not successful. For example, mother pigs of the Mong Cai breed were provided to ten farmers. They produced good piglets in the first year, thanks to the artificial insemination service from the university. However, without the support of researchers or students, their pigs could not get this service. Since no boar was kept in the village, the mother pig stopped producing piglets. Eventually, mother pigs stopped being kept in the village.

In general, the on-farm trials that used PDC yielded good results for the groups involved in pig production, fish raising, cassava production, and home garden and forestry production. For example, while only five farmers were using ensiled cassava leaves and roots to feed their pigs in 1998, 38 of them had started using that technique by 2003. Furthermore, the number of pigs raised in the community grew from 60 to 340. On its part, the fish-raising group, which grew from 12 farmers in 1998 to 54 in 2003, diversified its production from one type of fish to four, and increased productivity from about 4 tonnes per hectare at the beginning of the initiative to close to 15 at the end. Cassava production increased similarly, with more varieties now being grown for different uses, both domestic and commercial. Moreover, instead of growing cassava as a mono-crop, farmers now practise intercropping with beans and corn. They also apply contour lines to prevent soil erosion. However, one of the most drastic changes observed as part of this initiative occurred with home gardening. Indeed, while very few farmers used to work in their home gardens at the start, most of them now have home gardens linked to their production system (livestock, fishpond and home gardens). Animals, which used to wander around and destroy plants, are now under control in pens or have been fenced out, thus resulting in a reduction of conflicts in the village.

How can we work with the poor and the disadvantaged groups in a community?

This is a big challenge for any research or development group. Thus, it is important to find activities that really work for the poorer people. Usually, the poor and the disadvantaged do not participate in research or development programmes, and do not have access to the opportunities provided by outsiders or local leaders. For example, since extension workers or development officers want to achieve positive results from their demonstrations, they often work with the most advantaged people, who can demonstrate better results. Conversely, the capacity of the poor and the disadvantaged to apply production technologies that come from outside is limited. These challenges can be overcome if researchers or practitioners understand and follow the process of participatory development communication, while paying special attention to the poor. Second, farmers prefer to work in a group whose members experience similar conditions. In our case, similar-interest farmer groups proved to be a good way of working with the poor.

Furthermore, as part of this initiative, a PDC strategy was used to develop cooperation with the target people in the community. From that point on, action research was undertaken in order to meet the needs of farmers, and concentrated on identifying people, developing solutions and finding the best way to work with them.

Farmers also feel that participatory development communication has made them better understand the communication process between researchers and farmers. Because of their low education level, tools such as videos, cameras, leaflets, posters and role plays made them feel more confident when implementing initiatives aimed at improving their production.

Capacity-building with the poor

For most people in poor upland communities, the first need is food security, achieved by improving their production technology. However, in order to achieve sustainable development, capacity-building with local people is the next step after improving their immediate needs.

In any community, there are formal and informal organizations. Formal organizations are established to manage the community according to the current government system, such as the commune's people's committee, the farmers' association or the women's unionc. With regard to the development of upland communities, these organizations hold functions and responsibilities that have been assigned to them through formal regulations. In this case, the research team worked with the people's committee, the farmers' association and the women's union since these organizations have all established a good working relationship with the farmers. Meetings were organized with a view to empowering them and helping them improve their work.

Improving the capacity of leaders and commune organizations was the most important feature of building the communities' asset base. Participatory approaches were introduced and applied by these local organizations, thus encouraging people to define their own priorities and objectives, while making a contribution to the communities' plans and activities. Training and study visits were also organized to provide learning opportunities and to build people's confidence with regard to their relation to the outside world.

Training and cross-study visits for farmers were also organized, with an emphasis on the participation of the poor and women. The contents of training were developed to meet the practical problems of farmers. For example, before sowing their seeds, rice producers requested training. Given that they found it difficult to remember the material presented in a classroom setting, training sessions were organized in their fields. The fact that farmers learn better by doing became a communication challenge for researchers and practitioners alike since communication plans to be used in the training workshops had to be developed in order to better suit the farmers' conditions.

Given the existence of credit funds as part of this initiative, the women's union and the farmers' association were trained in managing small credit schemes and using credit money to invest in production. While 10 farmers in the commune initially participated in this scheme, after three years, 47 women had benefited from this credit fund.

Moreover, informal organizations such as groups of farmers were established to meet the demands of the poor. People usually have different interests and capacities in working. Each group included the farmers with the same set of interests. For example, groups were organized for farmers who preferred fish raising, pig production or home gardening. Within these groups, the farmers themselves established regulations on crucial issues such as membership and requests for financial or technical support.

Another form of asset-building consisted of working with the most disadvantaged in the community, particularly women. The fact is that the poor and women have a lot of potential for development. However, in most initiatives, the activities in which women and the poor are able to participate are usually quite limited. In this initiative, the team made special efforts to meet the poor in the community and give them various opportunities to become involved. There is no doubt that the design of a more appropriate communication strategy, coupled with the use of participatory approaches and tools, greatly facilitated the participation of the most disadvantaged groups within the community. Moreover, the fact that financial and technical support were made a priority also contributed to the successful participation of women and the poor, together with the support granted to their learning process.

A number of times during meetings held in the commune or in other districts, the leaders of the commune expressed the fact that the approach adopted by this initiative differed from previous ones. Indeed, giving farmers the opportunity to improve their understanding and working with them on what they are interested in has contributed to building their confidence. In the past, the ideas and priorities of local communities were largely ignored, while their local knowledge was simply not taken into consideration.

By way of conclusion, we can say that participatory development communication, as a new way of working with local people, improves the quality of the participation process. In order to use it properly, researchers must not just be open to acquiring new knowledge. They must also be willing to improve their skills in directions that may initially appear unclear to hard-core scientists. However, the benefits of their work can be greatly enhanced if these new forms of knowledge and skills are put to the service of the most disadvantaged.

Conserving Biodiversity in the Democratic Republic of the Congo: The Challenge of Participation

Pierre Mumbu

In places where biodiversity is under threat, governments often establish protected areas with a view to natural resource conservation. This intent is certainly laudable; but unless an effort is made to consult the people who have always lived from those resources, and unless they are involved in the decision-making process, conflicts are bound to arise. In this case, conservation efforts are likely to be in vain. The story of the Kahuzi-Biega National Park situated

in South Kivu in the eastern part of the Democratic Republic of the Congo (DRC) provides a prime illustration of the confrontational dynamics that are sparked when two contrasting approaches collide: that of the authorities seeking to protect a fragile ecosystem from over-exploitation, and that of indigenous people insisting on their ancestral rights over the territory. It also illustrates how participatory communication can help to find common ground between opposing views and generate cooperative action.

Located in the eastern part of the Democratic Republic of the Congo, the Kahuzi-Biega National Park (PNKB) is a major repository and sanctuary for biodiversity. It is home to 13 species of primates, 9 species of antelope and more than 400 bird species. Yet, since 1996 the park has figured on the list of World Heritage sites at risk because of intense human pressure on its resources. Until very recently, efforts to impose the prescribed ban on all exploitation of natural resources were met with stiff resistance from local people, especially the pygmies for whom the park's territory has been their natural homeland. The presence of indigenous people in the zone, who continue to pursue their traditional way of life, has tended to undermine conservation efforts, an effect exacerbated by uncontrolled poaching and the actions of armed militias.

The first efforts to protect the biological wealth of this zone were made a long time ago. As early as 1937, the Belgian colonial authorities established an integral reserve in a portion of this territory. In 1955, the protected zone was increased from 600 to 6000 square kilometres. It was officially made a national park in 1970 and was declared a United Nations Educational, Scientific and Cultural Organization (UNESCO) World Heritage Site in 1980.

The creation of the park was to serve two objectives: to preserve the typical mountain forest of Kivu and to protect a subspecies of gorilla that is found only in the forests of the eastern Congo. The park also constitutes the sole protected area in sub-Saharan Africa where there is continuity between lowland and highland forests, a circumstance that favours ecological exchanges between those two systems.

The park plays an important role, as well, in conserving the region's hydrography, rainfall and climatic equilibrium. Moreover, this natural forest has served to protect the water table upon which the inhabitants of the city of Bukavu and the villages located within the park depend for their water supply.

A complex set of issues

The portion of the park lying at higher altitudes is home to more than 0.5 million people. Only the original portion is uninhabited. But in the surrounding area, the density of human population is as high as 300 individuals per square kilometre. The natural forests in these areas have been virtually exterminated. The park has the appearance of a wooded island afloat in a sea of fields. Local people use the forest as a source of firewood and lumber, medicinal plants and foods of various kinds. The majority of the pygmies living in this section were expelled during the early 1970s and were resettled beyond the park boundaries, from where they continue to pursue their subsistence activities in the territory they previously occupied. It is from among the pygmies that the park recruits trackers for determining the location of animals and it is also the pygmies who provide the best guides for the poachers who are decimating the park's wildlife.

The 1975 expansion of the park brought 9000 villagers within its boundaries. At the same time, other people have been expelled. This is what happened to the Bananziga and the Mwanga-Isangi, who lost their traditional rights to these lands. In this area, the alternation of forest and village creates a mosaic within the park. Its inhabitants engage in extensive slash-and-burn agriculture, hunting, fishing and mineral digging, or small-scale mining, as well as many other activities necessary to their subsistence.

For the most part, it is the higher portion of the park that has been subject to demands for the return of lands. The traditional chiefs of this hierarchical society feel that they lost some of their power, which is essentially based on landholding, when a portion of their territory was incorporated within the park.

Tensions have been heightened by the fact that farmers have conspired with corrupt government officials to obtain and exploit fraudulent concessions within the park. When the farmers saw that these lands were to be taken away from them, they began to stir up opposition to the park's existence.

By the mid 1980s, it became clear that widespread local opposition to the very existence of the park was a serious obstacle to efforts to conserve its biodiversity. The authorities then recognized the need to hold discussions with the people concerned.

Problems raised by the local population

In the course of many meetings organized by the park authorities, local people described their view of the problem by saying that when the park was created, people were driven out of a territory that they had occupied for generations. Furthermore, when the park was expanded in 1975, people were neither consulted nor compensated. This fuelled resentment on the part of some village communities, who for this reason refused to recognize the park's existence. Previously, when the forest belonged to them, the indigenous pygmy population lived in harmony with its surroundings, and made rational use of the forest's natural resources.

Today the forest belongs to no one; it is government property. Consequently, the respect that ownership generates has vanished, and the traditional sound approach to the use of the forest has been replaced by a free-for-all to exploit its resources. There has also been a shift of mentality among some of the customary chiefs. The attentiveness that the chiefs traditionally bestowed upon the community as a whole has been replaced by a pronounced individualism. In some cases, the chiefs are now in the pay of rich and influential individuals who use them for commercial purposes. Great swaths of the territory have been ceded to rich farmers who are now growing monoculture commodities such as coffee, tea and quinine.

At the same time, the people who live around the park are poor and their numbers are growing. This exerts heavy pressure on the park since these people are in constant search for cultivable lands and staple necessities, such as wood for burning and for construction, wild game, mushrooms and medicinal plants.

There is also some confusion over the park boundaries, which have yet to be clearly defined. As a result, there have been some incursions of settlers back into the park, especially in its older portion. At this time, it would be difficult to force them out again without offering some form of compensation. Moreover, the migration of people with no 'forest smarts' into these wooded areas has caused considerable environmental degradation because they have no knowledge of traditional techniques of natural resource management.

In short, people see the park as a curse rather than a blessing. People struggling to eke out a living say that they are fed up with having their crops ravaged by the park's marauding elephants and they are tired of daily hassles with forest wardens.

Problems raised by the park authorities

For their part, the park's management felt that it was the lack of visible economic benefits that led people to challenge the park as an element of the region's economic development.

Although they do their best to respond to local people's demands, the park's managers must cope with a very heavy bureaucracy that prevents them from reacting promptly, and this is a further source of local discontent.

Finally, the villagers do not hold park staff in high esteem because of what they see as official harassment.

A development strategy for the area around the park

These discussions with local people revealed that, as with other national parks in the DRC, the Kahuzi-Biega National Park had always been run in a way that excluded local people from taking any part in its management. Not surprisingly, this exclusive model was found wanting. What was needed was an alternative model for managing and conserving the park's resources.

The authorities therefore undertook a number of development initiatives to help lift people out of their isolation and poverty. The intent was to foster activities that would, to some extent, compensate local people for the restrictions imposed on their use of the park's natural resources so that they would feel the park was making some tangible contribution to their development.

Beginning during the early 1990s, therefore, health and maternity centres were established, primary and secondary schools were built, and wells and water supply systems were installed, in collaboration with local people, while maintenance of infrastructure such as bridges and service roads was stepped up.

This new collaborative approach proved its worth during the two wars that tore South Kivu apart. In 1996 and in 1998, when the area was infested with armed militias, conservation staff dared not venture into the forest. Therefore, in places where there was good cooperation with the local people, such as in Kalonga and Nzovu, the people took it upon themselves to protect the park. On the other hand, where relations were less satisfactory, natural resources fell prey to massive and systematic destruction by the militias, by poachers and even by the local populace.

A 1996 survey found that around 36 per cent of people had a favourable opinion of the park's conservation, a surprisingly high percentage in light of the previous systematic opposition. However, an assessment conducted in 1999 found that the socio-economic activities supported by the development strategy, which was targeted at communities surrounding the park, were not really meeting the needs of local people, who were more interested in economic initiatives that would ensure their family livelihood.

In light of these findings and those of a more in-depth study of stakeholders in the park's natural resources, the authorities realized that they would have to strengthen collaboration with local people in order to achieve truly participatory management.

The participatory management strategy

Thus it was that, in October 2000, participatory management and communication mechanisms were established in certain villages on an experimental basis. These structures now offer a framework for cooperation, dialogue and decision-making. They are, in a sense, 'village parliaments', responsible for examining options and discussing and preparing a development plan for each of the villages bordering the park.

The members of these participatory management structures represent many viewpoints: public institutions, religious institutions, customary chiefs, leaders of local development initiatives, and other specific groups such as the pygmies, farmers and youth, as well as the forest wardens and women's associations. Each structure has a steering committee, a monitoring committee, a drafting committee and an anti-poaching committee. Since these participatory structures were set up, there has been a noticeable change in the kinds of initiatives put forward and greater attention is now paid to income-generating activities in order to meet locally expressed needs.

In addition to these participatory management structures, people have banded together in village social units to pursue collective self-help activities, rather than relying solely on individual initiatives. These units represent the backbone of management and support for locally inspired micro-projects.

Using the mass media

Along with the interpersonal communication on which the participatory structures are based, the park authorities have continued to resort to the mass media for the broader dissemination of information. Cooperative links have been established with a great many partners, who are provided with written and audiovisual materials that they can use in their activities. At the local level, use is also being made of the more traditional media, such as theatre, dance and popular songs.

Lessons learned from the process

The participatory management strategy has so far been confined to only about 10 per cent of the area surrounding the park because of security concerns generated by persistent armed confrontations. However, we can already draw some lessons from the process instituted to promote dialogue and cooperation between the park authorities and the neighbouring population.

The mode of communication that has proved itself most useful is that of interpersonal communication, where people address each other face to face at meetings or assemblies. The 'village parliaments' mentioned earlier are the prime example.

In the end, we may conclude that efforts to conserve biodiversity are more likely to succeed if policies and practices relating to management of the park's natural resources carry the stamp of consensus among stakeholders, rather than being imposed. In other words, free and open communication based on dialogue and mutual respect between the park authorities and the local people can overcome the resistance of those who initially felt excluded from decisions that were of vital concern to them.

To be sure, the communities concerned have yet to establish full 'ownership' over the development process, as they must if it is to be sustainable. In particular, once this experimental phase is over, it will be important to ensure that people are sufficiently equipped to continue with it and that they have the necessary means to carry on with the activities now begun. Otherwise, the crushing weight of poverty is likely to drive people back to helping themselves to the park's resources in the absence of viable options. This is a particular concern in the case of the pygmies, whose living conditions are still highly precarious. One of the greatest challenges, then, is to reinforce the existing income-generating activities and to expand their focus beyond sheer survival to embracing self-financing of local initiatives.

Yet, the long-term success of this strategy will depend in large measure upon strengthening dialogue between the authorities and the people, and maintaining attitudes of cooperation and collaboration on both sides. Here, again, the outlook is still uncertain, and it is not clear that the authorities are fully convinced that ongoing participatory communication is a key factor for future success.

Nevertheless, to date, the results are quite compelling. People in both places where this approach was tried out are playing a leading role in protecting the park's natural resources. Moreover, information now circulates more freely and an open climate of cooperation prevails among stakeholders. More concretely, since the participatory management strategy and the village economic units were established, goods taken from the park, such as game, bamboo and other forest products, are less in evidence in the village markets. Villagers are reporting poachers and have captured the occasional one themselves. They have even asked for portable telephones so that they can pursue poachers and smugglers more readily.

Yet, the sheer magnitude of the task is an obstacle to spreading the experiment to the entire local population. While there has been a considerable decline in poaching by villagers, there has been a sharp rise in the depredations of militias and other armed bands that have moved into the park. Against these illegal activities, the members of the participatory management structures are virtually powerless.

While awaiting a definitive end to the intermittent wars that have wracked South Kivu and that continue to prevent participatory communication and management from taking firm root in the villages that have tried it, another challenge is looming on the horizon: self-defence associations are springing up among the pygmies, who are claiming inalienable and unfettered aboriginal ownership rights over the park and its resources.

This situation, which would seem, at first glance, to contradict the participatory approach to managing natural resources, could be an important test for determining just how far participatory communication can contribute to preserving biodiversity without compromising the survival of cultural diversity.

The Word that Quenches Their Thirst: Rural Media and Participatory Development Communication in Burkina Faso

Souleymane Ouattara and Kadiatou Ouattara

Make water spring from the Earth and restore parched lands to greenery, and do this with empty hands? It can be done. The secret? Bring together people accustomed to working on their own, use the radio and share knowledge, including local know-how. This is a story that involves foresters, agronomists

and media people. But, above all, it is a story of the inhabitants of two villages (Nagreongo and Kriollo) in Burkina Faso and one (Kafela) in Mali. It tells of an experiment with action research that was marked by both successes and failures, but that provided plenty of lessons. The initiative was launched by the Journalists in Africa for Development (JADE) network, which was convinced that the rural media were missing a fine opportunity to support development because they lacked participatory resources and tools.

Barren soils, lack of water and accelerated deforestation are problems to which non-governmental organizations (NGOs) and government technical services are far from indifferent; but the solutions they try to implement are sometimes not very effective. In many cases, they pursue programmes already mapped out by their head offices, without any participation by the 'beneficiaries'. Moreover, even if they recognize the importance of community participation, they are not always well equipped to translate that concern into concrete action in the field.

It was to meet this gap that JADE undertook an experiment in participatory communication, an approach that involves local people in seeking local solutions to their own problems. The experiment was conducted in three villages, two of them in Burkina Faso and a third in southern Mali. Each of these villages faced natural resource problems that were holding back its development. Participatory communication served to rally stakeholders around solutions that they themselves had identified and agreed upon. This experiment produced lessons of three kinds, relating to the use of action research as a methodology, the conditions that must be present if interventions of this kind are to succeed and the choice of communication methods or tools.

Nagreongo: Too many women at the well!

In Mooré, one of the 60 languages spoken in Burkina Faso, the clearings where nothing grows are called *zippelé*. Exploited, abused, eroded by the rain and fierce wind, compacted by animals, the earth hardens into a sterile crust. The people of Nagreongo, a village in central Burkina, know all about the *zippelé*: for them it is a synonym for famine and thirst.

In March 2001, when the action research project was launched in Nagreongo, the village had 18,948 people and four wells. Not only were these water points insufficient, but two of the wells were out of service. With no management committee, maintenance work was random and haphazard. A famous healer lived in the village, and the sick would flock to consult him, at the same time imposing enormous demands for water and for medicinal plants. This village happened to be located within the broadcast range of a community radio station, Radio Yam Vénégré, which meant that people had access to information on farming techniques. They knew the history of their villages and they could listen to tales and legends.

Kriollo: When the water stretched as far as the eye could see

Further to the north in Burkina Faso is the village of Kriollo, an isolated settlement of some 2000 souls. When asked why people would establish a village in a place where there was no water for humans or animals, Dicko Issa Boureima, now around 50, recalled memories from his youth: 'I remember when I was still a teenager, perhaps 40 years ago, the water here was like a lake that extended all the way to Taaka, another village about 5km away.'

Today, farmland is in short supply, the quality of the soil is steadily collapsing and firewood is hard to find. In addition to these problems, the people are illiterate and women are not allowed to speak in public.

Kafela: Where a woman's worth is measured by the height of her woodpile

In Kafela, a Senoufo village of 510 people on the outskirts of the town of Sikasso in southern Mali, the forests are disappearing under heavy pressure from wood gatherers. The town of Sikasso generates tremendous demand for wood. To meet that demand, people are constantly hacking away at the forest, which is, in fact, their main source of income. Yet, the search for profits is not the only explanation. The community also carries some heavy social and cultural burdens. In these Senoufo lands, a wife's measure as a woman depends upon how much wood she can stack: a woman's social status can be gauged by the height of the woodpile in front of her house.

Action research as a methodology

Sharing the methodology

The first lesson from this experiment is that people have to agree upon the proposed initiative. But first, what is action research? Unfortunately, the term is not easily defined, especially when working with communities where the concept is foreign to the way in which they analyse problems and make decisions. This is why it is important, before undertaking any rural project of action research, to devote sufficient time to explaining the notion and to find the appropriate terms in the local language.

The weight of cultural heritage

Action research takes place against the backdrop of local traditions, which must be taken into account. For example, in many villages of West Africa, women are not allowed to speak in public. To get around this cultural prohibition, the project's fieldworkers set up homogeneous discussion groups consisting of women, men and youth. According to Awa Hamadou, a midwife in the village of Kriollo, 'Separating the groups meant that everyone could participate in the debate.' This was a first for a village that is heavily Muslim.

Today, the women are free to express themselves in public, in the presence of men, and on any topic they choose.

Voting and its limitations

The approach we used requires that the people identify and analyse their own problems, and set priorities among them on the basis of a direct vote. This is done by having each problem represented by a candidate. The approach helps to strengthen local democracy and the community's consensus-building capacity in a place where, by tradition, it was the elite who decided things for everyone.

This direct voting technique allows all the villagers to express themselves. With this rule, everyone can accept the majority choice. Yet, while the technique is certainly useful, it is far from perfect because people of the village are influenced in their vote by the views of the community leaders. One solution might be to establish private voting booths so that people cannot be swayed in their choice.

Putting things in context

We find that when communities have the chance to speak out, they will often raise matters that the project is not in a position to address. Faced with severe problems that threaten their very survival, people will often look for quick fixes, whereas a participatory approach places more emphasis on the search for lasting, consensus-based solutions. These two kinds of needs are not necessarily incompatible. The point is to help communities establish relations with other players who can provide the material support that they need. In the case of our three villages, the team was successful in this intermediation effort.

Choosing the players

Finally, the project's success will depend heavily upon the choice of players. They must not only have the required technical skills, but should be available and able to work in empathy with the communities, as well as with the technical partners.

Preconditions

Know your territory

The methodology requires a proper familiarity with the local setting as a precondition for subsequent phases. During a workshop held in Ziniaré in January 2001, extension workers in agriculture, animal husbandry and the environment explained their outreach techniques. Participants, including the local communicators, were then invited to comment on these presentations. They pointed out to the technicians the importance of the initial salutation phase, which is frequently cut short. They also recommended open discussions so that they too could propose solutions to the problems at hand. This exercise demonstrates the importance of 'knowing your territory', and here a contact person within the community can help. This person is usually known in Dioula, a language common to several ethnic groups in the region, as the *diatigitié* – roughly the 'host' – and it is he who will open your eyes and tell you what to do and what not to do while you are in the village. As the old proverb says: 'The stranger has big eyes but he can't see.'

Key players

During this initial phase, it is also important to identify the key players – that is, all those whose opinions count in making community decisions. In some cases, their influence will be such that they can stymie an initiative. The profile of the key player will vary from one region to another: he may be a healer, an imam, a customary chief or even an ordinary farmer.

Tools and methods for effective communication

The traditional means and vectors of communication

Each community will have its own means and vectors of communication. In most villages of West Africa, communication takes place at the church or mosque, at the well, in the market, in the village square, or at gathering points for young people such as kiosks, tea houses and popular celebrations. The project made good use of these traditional means of communication. In Nagreongo, the healer's representative, who was also the village information officer, had a megaphone that he used at meetings. In Kriollo, which is heavily Muslim, communication relied on the imams. In Kafela, the traditional chiefs transmitted information to the community. In all these areas, particularly in Burkina Faso, there is also a communication system based on the leaders of village groupings and on village administrative officers.

The project took advantage of these existing communication arrangements and was able to limit conflicts. But, at the same time, it ran the risk of becoming a prisoner of the system because discussions were heavily influenced by those in charge of circulating information, who thereby derived a certain power. Clearly, a proliferation of communication channels and tools is no guarantee that everyone (particularly the women and youth) will be involved in decisions that commit the community. The challenge, then, is to use existing means and vectors judiciously and to propose ways of improving them.

Facilitation

For the project coordinator, facilitation means encouraging people to pool their efforts and their skills in order to resolve local problems through natural resource management. The multidisciplinary nature of the group thus constituted appealed to people in the participating villages. One farmer explained:

> *In the past, visits to our village by members of this or that regional committee or agency were random and disorganized. They took up a lot of our time and people were tired of going to meetings, although we never said that to the experts. They did not even answer all our questions. Now that we have this group, we are finding answers to all our concerns and we are also saving time.*

As to the agencies, they are not only getting firsthand experience of local people's problems, but are now taking them into account in their follow-up activities. This new partnership goes well beyond the old working relationships that were based on institutional considerations and made no allowance for a common approach.

Using local languages

Using local languages is essential for sharing information, thinking and outcomes with all partners in the process. The local relay points, rooted in their home communities, are already using the national languages effectively to deal with communities in the course of their daily work. Radio stations, for example, are broadcasting in Mooré and Fulfuldé at Ziniaré and Dori, and in Bambara and Senoufo at Sikasso. A newsletter, called *Kuma*, has also been translated into Mooré and Dioula, to an enthusiastic community reception. Although people in these communities were literate, they were not always able to find publications in local languages, and were thus unable to keep up their reading skills. Kuma is now helping to fill this void.

Local languages offer many advantages; but translating concepts such as action research, participatory communication and research project into these tongues can be a complicated affair. A great deal of time was spent discussing translations within the team in order to test out new expressions before adopting them and disseminating them locally.

Finally, the diversity of languages is such that research team members will never be able to master all of them. During our meetings, we therefore had to arrange for translation from French into the local tongues and back, and this in effect doubled the length of time that would normally be devoted to such meetings.

Discussion groups

Discussion groups make it possible to compare different ways of addressing a problem. They can also be used to uncover conflicts of interest and to find solutions that will be fair to all parties involved. The advantage of these groups is that everyone has a chance to speak, including women and youth who are normally shy about expressing themselves in public.

The limitations of radio

The research team launched its activities by radio, which is a powerful information and communication tool, especially in rural settings where the oral tradition dominates. Editorial committees were established in each of the three zones (Sikasso, Ziniaré and Dori, where the project's partner radio stations were based) in order to produce regular information programmes. These committees operated in the conventional editorial mould, with the one difference that in addition to radio producers, they also included representatives from the communities and from development agencies.

The synergy that emerged among all these players enriched the content of the 'magazines' (reports, surveys and technical data) and represented a real experiment in partnership. Unfortunately, the development officials often steered the debate towards their own particular concerns. The village communities, the prime audience for the message, were not sufficiently involved in choosing the information topics. In order to overcome this constraint, we set up a network of local communicators in Dori and strengthened those already existing in Sikasso and Ziniaré. The local communicators were responsible for identifying topics in advance, preparing field trips for the radio team, taking part in the broadcasts themselves and providing feedback to the stations.

The workshops

The workshop meetings served a vital function throughout the participatory communication experiment. They allowed people to expand their normal circle of relationships and to replace an ethos of competition with cooperation.

Conclusions

Whenever communication problems existed within the communities, synergy among the stakeholders, through group discussions that identified and analysed problems, served to establish an open dialogue and to make decisions that would commit the entire community. In this way the community was able to find home-grown, consensus-based solutions to the water shortage, to the disappearance of the forest and to the collapse of soil fertility. This open approach, and particularly the participatory dimension of activities, also tended to temper the climate of competition that often exists among development workers in the field.

The communities who threw themselves most enthusiastically into the project believe that action research and participatory communication are appropriate and are a readily usable means of addressing their development problems, instead of waiting for unpredictable financing.

Growing Bananas in Uganda: Reaping the Fruit of Participatory Development Communication

Nora Naiboka Odoi[1]

Banana is one of the most important crops in Uganda, and in many homes, especially in the central part of the country, it is the staple food. Many small-scale farmers derive both income and food from the banana crop. But since the 1970s, many of them have been experiencing decreased farm yields as a

result of declining soil fertility, pests, diseases and socio-economic problems. In this changing context, traditional practices have proved insufficient to cope with the new challenges. On their part, researchers have come up with technologies that could be of benefit to farmers. But the latter have proved reluctant to integrate these new technologies within their practices, despite researchers' attempts to disseminate them.

This scenario has led some agricultural researchers to question their way of communicating with farmers and to experiment with more participatory approaches. This story is about farmers who have been so successful in solving their own problems that they have become teachers to other farmers and development communicators in their own right.

Despite the growing hardship and difficulties associated with their traditional agricultural activities, farmers tend to overlook the new resource management practices proposed by researchers. When they do use them, it is only during the period that researchers are with them. Therefore, agricultural research findings have not been integrated in a sustainable manner, even when the initiatives bear positive results.

One of the reasons for these failures points towards the fact that researchers and extension service providers have largely relied on top-down dissemination methodologies in which farmers are not involved in decision-making regarding which technology to implement in their gardens. They do not own the natural resource management initiatives being tried out in their own gardens. When researchers visit the villages, some farmers have been known to show the researchers two plots of gardens: their own and that of the researchers. 'This is our garden, and this one is yours', the farmers are often heard to say, the second garden being the one where farmers are putting into practice the technologies recommended by the researchers.

For the past few years, researchers have been on the lookout for a methodology that may result in farmers owning the natural resource management research initiatives so that they are able to implement them in their own banana plots in a sustainable manner. In this sense, participatory development communication (PDC), which is a relatively new concept in Uganda, promises positive results regarding information-sharing between farmers and researchers. During recent years, several organizations have started to implement participatory methods, albeit with varying degrees of local people's participation. Uganda's National Agricultural Research Organization (NARO) implemented one such initiative through its National Banana Research Programme. The two-year research initiative called Communication Among Banana Growers for Improvement of Soil and Water Management aimed at developing a two-way communication model suitable for a better flow of information between researchers and banana growers, at enhancing farmers' participation in experimenting with different banana improvement technologies, and at fostering farmer-to-farmer training. The study used PDC as a tool to foster active participation of the local community in identifying natural resource management problems in banana gardens, as well as their causes and solutions.

Initiating dialogue

After a series of in-house planning meetings, a team of researchers – comprised of a socio-economist, a communication specialist and counterpart, a soils expert and a specialist of the site – toured several banana-growing districts. They were in search of a suitable site that would host the natural resource management through PDC initiative. The team zeroed on Uganda's south-western district of Rakai, most particularly the sub-county of Ddwaniro. This choice was motivated by the fact that Ddwaniro is among the leading banana-growing areas in the district; it also has soil management problems as a leading farming constraint, a relatively good road network and is inhabited by hard-working farmers.

The research team first held a series of consultative meetings with district and sub-county officials, opinion leaders and extension agencies working in the sub-county before holding any meetings with farmers. This step facilitated the introduction of the initiative to the named groups, in addition to lobbying for their support. Farmers were encouraged to form farmers' groups, whose representatives later participated in the PDC initiative.

The researchers facilitated the identification and prioritization of natural resource management problems by the farmer representatives. In order to do this, they divided themselves into three groups, in line with problems pertaining to soil fertility, soil erosion and soil moisture retention. The farmers then defined their communication objectives and needs regarding the identified problems, the activities that could be undertaken to alleviate the problems and the communication tools that could assist them in sharing their new knowledge with their original farmer groups.

The researchers discovered that some farmers had extensive indigenous knowledge related to the three natural resource management concerns, but that it required validation. Moreover, farmers did not have a forum to share information with each other, hence the need for communication tools.

Horizontal communication processes

The researchers facilitated participating farmers in visiting other farmers who were already using more appropriate natural resource management practices. Following the visits and the discussions with other farmers, the participating farmers were more convinced than before of the new technologies' benefits. The researchers then facilitated the implementation, by the farmers themselves, of proper natural resource management in their own banana plots.

The farmers were amazed at the results of the new practices. Plots of land that they had previously abandoned, saying that they could never be productive, were now yielding good banana fruits. Their banana plots looked healthier, and they yielded better quality bananas in greater quantities than before. They were now able to sell some of the bananas for cash, while remaining at the same time with enough left over as food to eat in their homes. Banana farmers have a practice of selling bananas when the banana fruits are still on the parent plant. The bargaining process between the seller and the buyer often takes place within the banana plot while the plant is still intact with its fruit, a practice believed to be due to the perishable nature of the banana fruit. After implementing proper natural resource management, the participating farmers said that their sales fetched more money from buyers because the amount of invested labour was evident in the banana plot. A well-managed banana plot made people more inclined to buy the product, like a proper packaging of the banana fruit.

The farmers were now the proud owners of well looked after and high-yielding banana plots. Unlike before, farmers were confident to show their banana plots to other farmers and visiting dignitaries in their community. Local leaders started using the banana plots of the practising farmers as showpieces. Natural resource management through PDC raised the prestige of the practising farmers. Several of them became leaders in their communities. Some of them even managed to contest successfully for political leadership in their community.

But the practising farmers never forgot the fact that they were only representatives of other farmers in their local communities. After they had mastered the natural resource management technologies, they expressed the desire to share their new knowledge with other farmers and the farmers they had represented in the initiative. They recognized that they had to use communication tools that had the capacity to illustrate how to implement natural resource management technologies. They chose to use video, photographs, posters and brochures on soil fertility, soil erosion and soil moisture retention.

Looking back on the experience

The concept of participatory development communication was new to researchers. This resulted in the loss of valuable time since several activities had to be redone. The biggest challenge happened at the production stage, when both researchers and farmers were new to using the communication tools. Eventually, professionals (a graphic artist, an illustrator and a cameraman) had to be called upon in order to successfully produce the materials.

Although farmers wished to practise the new natural resource management techniques, they indicated constraints of money, labour and agricultural tools. Some of these problems were institutional in nature and could not be solved by communication alone. Thus, the researchers and the farmers themselves sought partnerships with other organizations that could help them to solve these problems.

The production process took longer than intended. This was because farmers first had to learn how to use the communication tools before focusing on natural resource management. The learning process was not straightforward: the farmers had neither handled still cameras nor video cameras before. Consequently, there was a period during which farmers were only building their confidence to handle the equipment, which they considered to be too expensive for them to replace in case of damage. Eventually, they became confident enough to handle the cameras. There followed a process of perfecting the capture of usable photographs. Many films returned half pictures: pictures of children with no heads or pictures of people with half faces.

Farmers discovered on their own what type of photographs could successfully communicate information regarding proper natural resource management. Thereafter, they set about trying to acquire those pictures. Unintentionally, the process of taking and retaking photographs deeply ingrained the subjects in the farmers' minds. Without realizing it, the farmers were learning more about proper natural resource management while their focus was on taking pictures of the technology used.

After the communication materials were produced, participating farmers showed them to other members of the community, who made comments about them. One such comment required a change to be made in how a good banana mat looks. The original picture showed banana mats with single banana plants; but it was pointed out that a recommended banana mat should have at least three banana plants: 'a mother, daughter and granddaughter'. A change was also made in the video. This followed a recommendation that a song should be included in the video. On the whole, the other members of the community appreciated and understood the communication materials.

Monitoring and partnership formulation proved to be another big challenge to the researchers. They only realized at the end of the phase that these two steps should commence at the beginning of the initiative and continue throughout the process.

By the close of the first phase of the Communication Among Banana Growers for Improvement of Soil and Water Management initiative, farmers had appreciated the power of belonging to a group. As a result, they founded an association called the Ddwaniro Integrated Farmers' Association, through which they hope to search for, access and share relevant information and services about common community problems. This is the origin of a two-way communication forum between farmers and other stakeholders, including service providers and researchers. Such communication will bridge the gap between researchers and farmers, and supplement the inadequate extension services.

The farmers can now tackle their own community problems instead of waiting for external assistance. They are proactive instead of passive observers of community problems. They do not fear approaching service providers regarding their community concerns. In addition, other people find it easier to assist farmers when they are in groups. But unlike other groups, the group that formed after farmers had undertaken a common activity is stronger and promises more sustainability and more focus. It is a secondary group, unlike other groups that are only primary groups. A secondary group has more chance of implementing farmer-to-farmer teaching because there is already another group of farmers waiting to get information from it. This illustrates a multiplier effect of the PDC methodology.

The researchers and other stakeholders who participated in the Ddwaniro initiative are now convinced of the power of participatory development communication in implementing natural resource management initiatives together with farmers. They have begun to incorporate PDC within their research initiatives. Farmers now master the three methods of natural resource management to the point where they are now confident enough to share the information with other farmers. The farmers of Ddwaniro have begun sharing their experiences with other farmer groups within Rakai and other districts. This is a result of other farmers recognizing the positive results of the Ddwaniro initiative, and asking for similar initiatives to be introduced in their own areas. On their part, women are no longer as shy as before. Men have seen the benefit of their wives taking part in the PDC ventures, and they now readily allow them to attend meetings.

The way ahead

Any initiative begins with a proposal. For a long time, this has been done by researchers alone. Recently, researchers have discovered that it is possible to write a research proposal together with other participating stakeholders, such as farmers. Once the research proposal has been approved, the researchers, together with other stakeholders, can agree on the actual implementation of activities, based on the participatory development communication methodology. After having started with problem diagnosis and prioritization, the process then moves on to identifying solutions and implementing them, followed by evaluation and report writing. Indeed, PDC is most successful when it is used throughout an initiative, and not only at certain stages.

When looking ahead, it appears seems that PDC is a skill that farmers can transfer to other areas of their lives. In this case, farmers have used it to solve natural resource management problems. They could also use it to solve medical, social, economic and political problems. Farmers have the potential to undertake many initiatives on their own; but this potential often needs to be re-awakened before they can make use of it. After an issue has been discussed, farmers gradually realize that they can undertake certain tasks that they initially did not think about, or that they initially thought they were not capable of implementing. Whereas some farmers may recognize on their own that PDC can be used in different contexts, there may be a need to facilitate other farmers in recognizing the possibility of using PDC in varying spheres of their lives.

For some, participatory development communication may initially appear too expensive, considering the time, staff, travel, money and other logistics required. In the case described here, it took some time before farmers and scientists began appreciating its power and benefits. But after implementing the first phase of this initiative, the benefits seem to have dwarfed the initial costs. The 60 farmers who participated in the first phase have multiplied the efforts of the one agricultural extension worker in their area about 30 times by teaching other farmers in their immediate neighbourhood and beyond.

The other farmers are extremely happy with the farmer teachers because they are easily accessible. Moreover, they understand their problems better than outside agricultural experts. The farmer teachers give hope to other farmers by showing them that the natural resource management issues they are teaching can be easily implemented. Other development agents are now planning to hire them to teach other farmers in agricultural-related initiatives. Indeed, these farmers have gained so much confidence that they have begun filing radio items related to their natural resources management teachings to local FM radio stations in order to extend their accomplishments and knowledge beyond their own communities.

Note

1   This chapter was written in consultation with Wilberforce Tushemereirwe, Drake N. Mubiru, Carol N. Nankinga, Dezi Ngambeki, Moses Buregyeya, Enoch Lwabulanga and Esther Lwanga.

Giving West African Women a Voice in Natural Resource Management and Policies

Rosalie Ouoba

In many West African cultures, education and cultural norms do not encourage women to speak out in public. Yet, women constitute around 53 per cent of the population and 80 per cent of them live in the countryside. They account for 90 per cent of agricultural labour and they do most of the transporting and processing of farm and forestry products. Prohibited from speaking out in their communities, women are thus denied the right to

participate in decisions about the management of natural resources, of which they are the primary users. The Union of Rural Women of West Africa and Chad (UFROAT) decided to tackle the situation in a new way by enlisting women from six countries[1] to participate in drawing up an action plan:

> *My name is Sali Fofana. I come from Dafinso, a little village in western Burkina Faso. My daily life revolves around the search for water, which takes longer and longer to find. Our village has only two pumps for 1800 people. One of the pumps has broken down, and so we have to spend half the day lining up in the heat of the sun to get our supply of water. For us, water means life, as you will hear people say in the Sahel. We have to find enough for our people and our animals. I am so overwhelmed that I have to send my daughter out for water. That means she does not attend school regularly, and she's not doing very well in her class work; but what can I do? Not only that, but the marsh has been dry since January. It's as if nature herself has abandoned us to our fate.*

Sali Fofana is not a real person; but what she describes is very real for millions of African women who feel directly the impact that environmental collapse is having on their living conditions. Her words could be those of any of these women. They sum up the experience that a great many rural women have related in the course of an initiative to make their voices heard in decisions about natural resource management policy in their region.

With the accelerating disappearance of vegetation cover from countries of the Sahel, not only is water becoming scarcer, but people have to travel ever greater distances for the firewood that is their chief source of domestic fuel. The situation falls heaviest upon women such as Sali Fofana, whose stories often go something like this:

> *In the past, when I needed wood for the kitchen, all I had to do was go into the field next to the house. Today, wood is scarce and I have to go a lot further to find it, or I have to buy it. Since this is now a lucrative business and the distances are great, it's the men, with their trucks and carts, who go out for the wood and then they sell it at ever steeper prices.*

It was no surprise that the 1992 Earth Summit in Rio de Janeiro emphasized the role of women in preserving a viable and stable environment, while stressing the need to reduce extreme poverty, which is closely linked to environmental degradation.

Given their key role, rural women should be a great asset in developing natural resource management policies. This was precisely the objective of an initiative to enlist rural women from six West African countries in drawing up an action plan for natural resource management in a process that was participatory from beginning to end. With the implementation of that plan, they would, in turn, become a force for pressure and a source for ideas concerning national and regional policies. Moreover, the process sought to demonstrate that participatory communication and, in particular, the exchange and sharing of knowledge can help rural women to analyse the problems they face and seek solutions by drawing upon the knowledge, experience and resources that they possess.

Methodology

The initiative has involved two stages to date. Initially, workshops were organized in each of the countries to help the women prepare the materials that would be used in the action plan. The second stage was to formulate the plan. In 2005, there was a third stage, which saw a regional workshop where women from the six participating countries validated the plan.

In methodological terms, the workshops were designed to let the women express themselves by defining and analysing their own situation and coming up with their own solutions that will take their specific circumstances into account. Talking is the main tool in these workshops. The initiative also exhibited the following characteristics.

Recognizing the diversity of rural women's concerns about ecology and farming

Although the degradation of natural resources is a generalized phenomenon, its severity and the constraints that it generates vary from one agro-ecological zone to the next. In order to reflect this diversity of problems, representatives were selected from every agro-ecological zone in each country, and were invited to express themselves in their own language.

Facilitating information on key concepts

These workshops employed two key concepts: natural resource management and action research. Experts from the environment and natural resource ministries and resource persons in different fields were asked to contribute to these workshops. After listening to all the viewpoints expressed by the women, they offered some explanations on these questions and on the national policies to which they related.

The organizers had to draw heavily on their experience in order to achieve the workshops' objectives, given the women's low level of education.

Participatory diagnosis of constraints in each zone

The women were divided into groups according to their home region. Focusing on the three components of natural resources (land, vegetation and water), each group described and analysed the status of natural resources in each agro-economic zone, as well as the role of women in resource management and the difficulties that they encountered in performing that role. Next, the groups compiled an inventory of possible solutions and the needs facing women under each of these components.

Interaction among the groups

The working groups reported their results to the plenary sessions, where the women debated both the difficulties they faced in common and those specific to certain zones. They then discussed and prioritized the activities to be included in the action plan.

Field visits

Visits were arranged to meet with women's associations working in the field of natural resource management. This brought the participants into contact with other women's associations engaged in activities similar to their own. The visits provided the opportunity for very useful exchanges between participants in the programme and the associations visited. They also laid the basis for an ongoing relationship among women from different agro-ecological zones.

Some eloquent stories

Throughout the process, participants discussed the impact of gender inequality on rural women's lives and on their efforts to preserve natural resources. The story recounted by Sali Fofana provided eloquent testimony of how working and living conditions for female farmers have deteriorated over the years:

> *My field is barely half a hectare. It's a long way from the village and from my husband's field. It's an old field and it's not very productive. I have to get up at the crack of dawn so I can get my own fieldwork done before I have to go and work on my husband's field. I have no one to help me and I don't have any modern tools or inputs, such as a plough or mineral fertilizers, and so I put a lot of hard work into my field for a very meagre return. This year, for example, I harvested only two bushels of peanuts. Of course, I also grew some okra, some sorrel and some eggplants to perk up the family diet.*
>
> *A few years ago, like all the women in my group, I learned how to produce organic fertilizer. But I don't make any use of it; if I enrich my field I'm afraid it will be taken away from me. It's the chief who gave me my field. In our village, women don't have the right to own land. This means they can't plant trees on the lands that are given to them, nor can they take any major steps to protect the soil.*
>
> *I like to gather shea nuts and seeds from the locust tree [néré] to use in the kitchen and to earn a little money on the side. Nobody grows these things. You find them in the fields and in the bush. Today the bush is steadily retreating and it's becoming harder to find these fruits. We can no longer get any soumbala [a kind of mustard from the néré, used for seasoning] or shea butter, and now we have to buy our shea nuts and locust seeds from the men. This is really serious because we're not*

allowed to go into the fields to gather them. Since these wild fruits are worth money, the men keep the harvesting of them to themselves. People are so greedy that they will even take fruits before they are mature and then they use chemicals to speed up the ripening, something that can be hazardous to your health.

The local organization as a talk forum

Yet, despite all these difficulties and setbacks, women are not giving up. In some cases, their resolve has led to the creation of groups that are now the main forums for sharing information and talking about their daily lives. These groups have produced leaders who are now invited to represent them nationally and internationally, as was the case with this initiative. As Sali Fofana puts it, things unfold as follows:

As you can see, I'm tied up with problems. But thank God I'm a fighter. So I got the women in my village to group together to try to resolve our problems. After all, we want a bit of happiness too. My village sisters asked me to head up the group. They chose me because I get along well with everyone, I lived abroad in Ivory Coast and I went to school as far as the intermediate level.

In the invitation letter, I was asked to organize a meeting with the women and the men to discuss the status of the environment in our village and to talk about the place and role of women. This was important because we have to consider all points of view if we are to take decisions that will help the whole village. If we, women, just got together alone to analyse things, that would be all very well; but the men might feel left out and they could put roadblocks in the way of our project. As it is now, they are our allies.

Kaya: Women and experts

Following this local initiative, a national workshop was held in each of the six countries, bringing together women from several regions. In Burkina, participants met at Kaya, north of the capital. This workshop was intended to pool the women's experiences and observations and wrap them up into a coherent action plan. For Sali Fofana, as for all the rural women who took part, the opportunity to express the viewpoint of her village and to compare notes with the other women was an invaluable experience. And when their knowledge and their observations were validated by experts in various disciplines, this served to reinforce the women's perception of the importance of their role within their communities. For all the Sali Fofanas who had invested time and effort in the process, this was a moment of great pride:

The great day had arrived; I was off to Kaya. There, I met women from the four corners of Burkina. Each of us spoke about her village, about rainfall patterns and their impact on the bush, on water, on people and on animals. Each of us talked, as well, about her role as a woman in exploiting and managing the bush, the water and the land. We realized that a woman's situation is just about the same all over the country. There were natural resource management experts present and they gave us some new information that we used to analyse the situation in depth. They admitted that what we were saying was right on the mark. Like other countries in the Sahel, Burkina has been suffering a prolonged drought since 1969 and rainfall has dropped sharply. This has had a severe impact on vegetation, on soils and on farming systems, leading to an explosion in the area under cultivation and a drop in fertility and in the availability of water. While national guidelines set a standard of one water point for every 500 people, the real figure in the countryside can be multiplied by five or six times, and around 57 per cent of households get their water from undrinkable sources.

We found some solutions and we proposed some actions. Take the fruits that we use for our condiments, for example. We said that men and women have to work together, within the home and in the village, to appreciate the usefulness of these fruits in feeding the family and to enforce village regulations that used to prohibit people from taking the fruits too early. The authorities used to decree a specific day when everyone could go out and gather the fruits. We waited until they were ripe and no one could take them before. That regulation has to be brought back into force in the village. Then everyone will have a little, and that will prevent the fruits from being gathered when they are still green and unusable.

We want to be involved in selecting the species for reforestation. Since the locust and the shea don't yield as much anymore, we need to plant more of them, as well as fruit trees that we women can use to feed and care for our family and to earn a little money. We hear that shea butter brings a lot of money today because the Europeans like it and they buy a great deal of it. We women know how to produce the butter, we are trained, and we can look after the trees and make a high-quality product that will bring even more money.

The locust tree produces soumbala, a spice that every woman knows how to prepare and that gives a good taste to our sauces. Since the locust no longer does well, soumbala is expensive. It is being displaced by the Maggi cube, which is widely advertised. People say that's why so many people are dragging along with high blood pressure. We want to bring back our native plants that allow us to eat and to earn a little money.

When it comes to water, we have asked to be involved in selecting well sites and to be given some decision-making positions on the water point management committees. We're also going to demand more water points in the villages so that water will be more readily accessible.

As to the women's fields, we want the ones they give us as our personal plots to be as close as possible to the family field, and we want to be able to use our husbands' tools and materials to cultivate our fields. We also want to see discussion in the village about the importance of the women's fields in terms of family food security. More and more, our families depend on the produce from our fields to tide them over until the next harvest. We have problems in producing the compost that the extension workers have taught us to make, and sometimes we have to steal it and put it on our fields. If we had more compost, we could cultivate our plots more intensively and produce more food. With all the solutions proposed, we discussed their feasibility and we selected the best ones.

Lessons learned from the experiment

Involve the women in data gathering

To promote community participation, the women were asked to do some advance preparation before the workshops whenever they could. In some cases, this preparation involved the entire community, while in others it was limited to the women's associations in their specific configurations. These discussions within the community not only served to inform the community about the important role that women play, but sensitized people to the problems women face in carrying out their tasks. The results formed the basis for activities in the workshop groups.

Have participants analyse the data

The approach used was shown to be very effective because it respected the ability of rural women to define their place and their role, as well as the constraints they face in managing natural resources. In their analysis, they took into account the pattern of distribution of household tasks while stressing the need for dialogue between men and women in order to re-establish a degree of balance and equity. They identified their needs very clearly in terms of natural resource management, and they put forward solutions that took account of the whole community.

The methodological approach

Allowing rural women to speak and appreciating the value of their knowledge can instil confidence in this social group, which suffers daily from injustice and inequalities, and can encourage them in constructing a vision of society.

The right to speak is very important in the life of every person, of every social group and of society as a whole. It lets individuals or groups express what they are, what they experience and what they feel. It also allows them to affirm what they know, what they can do, what they think and what they want. But, above all, speaking allows individuals or groups to organize themselves in order to improve the situation or achieve the goal by entering into communication with other people or other groups to enrich the social debate. Finally, speaking allows traditionally marginalized individuals or groups to participate in taking more relevant and equitable decisions for sustainable development.

In short, speaking is a fundamental human right. Yet, given the balance of forces within societies, this right is denied to some people, particularly women.

Despite the real progress that has been made during recent years, the overwhelming majority of rural women in West Africa still have no voice. Yet, when they are given the opportunity, they can define and analyse better than anyone the situations that affect them, provided that the facilitator respects their pace and gives free rein to their capacities for imagination and expression. The right to speak must mean the right to express oneself; but it also means listening and communicating, for it is through communication that new visions are constructed. Thus, if rural women express themselves and compare their views with those of other groups (through dialogue, exchange of knowledge and study visits), they will be able to decide the best actions to take as a group to secure sustainable development for their communities. At first, they will talk mainly about 'practical needs'; but they will very quickly move on to 'strategic interests' (transforming unequal relations between men and women, or between rich and poor). When they talk about the actions they want to take to meet their needs, women always demonstrate a concern to discuss them with the men and with their communities. In the end, most of the actions they select are things that they can do either by themselves or in conjunction with the partners working with them.

Problems encountered

The approach taken in the workshops in the six countries, while empowering, was beset by problems relating to the limitations, both intrinsic and extrinsic, that are imposed on women.

The diversity of languages

In some of the workshops, there were nearly as many languages as participants, and we had to find local facilitators who could serve as interpreters. Apart from Mali, where nearly all of the women in the workshop spoke Bambara, proceedings had to be conducted in at least three languages. It took a lot of work to prepare the interpreters to give translations that would respect all the subtleties. It was very frustrating to have a woman recount her experiences in her own tongue and with great emotion, and then to hear a translation that would render not half the story and nothing of the emotion. We can imagine that rural women have the same feeling when debates are conducted in French, and when they have to wait for the translation in order to understand proceedings and offer their own viewpoint.

Moreover, the time it takes to go through this exercise can discourage participants, who would rather engage directly with the audience.

Socio-cultural constraints

By tradition, women do not participate in the discussion of community problems. They do not express their views in public, unless they are mandated to do so by a group. Most of the women attending the workshops, even those representing associations, were unaccustomed to speaking publicly and their lack of experience made them uneasy about expressing themselves spontaneously.

In fact, given the social division of labour, the many domestic tasks that fall to women leave them little time to attend meetings. Thus, they never acquire the habit of participating or of speaking in front of men.

Religion (or its interpretation)

Because of their religion, some women are not allowed to frequent places where men are present. Their daily life is cloistered and they take little part in associations of the kind that allow women to open up and let down their guard. Those who get the chance to attend workshops of this kind often have trouble expressing their viewpoints clearly.

Women's illiteracy

The fact that most rural women are illiterate is an enormous constraint on their ability to interact with others. Among the group (from six countries), only perhaps one woman in four could read and write (either in French or in her native tongue). Despite the effort to provide illiterate adults with audiovisual tools to help them receive and exchange information, we must recognize that the rural world remains largely cut off from the exchanges of information that are essential to thinking about development.

Lack of education

Analysing a situation or problem requires capacities for assessment, reflection and reference to other experience or knowledge. Because they are confined in their environment and have no access to information that might deepen their understanding, rural women have trouble analysing a problem in depth, identifying all its causes and ranking them in order to find all the possible solutions. Their analysis, however relevant it may be, is limited. They jump too readily from a problem to the solution without appreciating all the underlying causes. Rural women need to broaden their horizons and interact with other people.

Some results from the workshops

During the national workshops, the women analysed the natural resource situation in their communities, listed their concerns and their needs relating to natural resources, and identified the main actions required to meet those needs.

The women's analysis of their situation showed that, to varying degrees, natural resources in all their countries are deteriorating at an unprecedented pace that threatens people's very survival. This sparks competition that throws into conflict the various components of a given community, or of two or more communities who share these resources. It is most often the weakest members of the community who are excluded from managing these resources, and women fall in this category.

These workshops also revealed the women's awareness that improving the conditions under which they work is a prerequisite to any solutions for preserving natural resources. They want to change their relationship with their communities through dialogue so that their family members and friends will appreciate their contribution differently. Despite differences from one agro-ecological zone to another, some common features stand out clearly, such as the need to overcome inequalities of access to land and property rights. Land, water and vegetation are at the very centre of women's lives, and they will be the first to try to preserve those assets if they are given the means.

Participants in this initiative wanted to produce an action plan that would be an instrument for reinforcing their capacities to analyse, to compare and to propose solutions so that they could constitute a critical mass for reversing current trends. In the end, they wanted to develop collective strength so that they could negotiate and influence decision-makers in order to change their own living conditions and those of their communities.

A first outline of the action plan became available in September 2004. To turn this into a real tool for members of the UFROAT rural women's associations, it will now have to be validated. A regional workshop with all the delegates from the six countries should serve to deepen the analysis and confirm or reject some of the proposed strategies.

By way of conclusion, we may say that overcoming the many obstacles that prevent rural women from speaking out means establishing a framework and conditions within which women can affirm themselves and their contribution to society, where they can broaden their horizons and encourage society's awareness of the importance of seeking and listening to their views and taking them into consideration at all levels of debate and decision-making bodies.

In a sense, facilitating dialogue among rural women, with other members of their communities and with development players, means unleashing at least 50 per cent of the region's potential energy and putting it to work for development.

Note

1   These six countries comprised Benin, Burkina Faso, Mali, Niger, Togo and Chad.

Water: A Source of Conflict, a Source of Social Cohesion in Burkina Faso

Karidia Sanon and Souleymane Ouattara

Women glaring daggers at each other, husbands ready to come to blows, and rivalries between former villagers who have moved to the city, between ethnic groups and between adherents of different religions. Can anyone imagine a more explosive mixture? Yet, this is the time bomb that the researchers of GUCRE (Management of Conflicting Uses of Water Resources project)[1] have set out to dismantle by facilitating peaceful access to water for everyone in the Nakambé Basin of Burkina Faso. They are succeeding in their

wager. They have made social cohesion their battle cry, and they are using participatory chat sessions, to which they have added other tools, such as forum theatre, video, radio and posters. Here is a story from Silmiougou, a village in the heart of Burkina.

This is not exactly the forested West, far from it. But neither is it the Sahel. Here at Silmiougou, acacias, ana trees and mangoes grow side by side with thorn bushes, the precursors of the sparse vegetation of the great North. Silmiougou, a village in the heart of Burkina Faso, still dreams of its verdant past. Listening to the village elders tell the story of this settlement, we can readily see how the environment has deteriorated. According to oral tradition, it was Peul herders who originally occupied this area. Under constant attack from rustlers stealing their flocks, the Peuls complained to the king of the Mossi, who sent them his stout warriors for their protection. Little by little, the Mossi protectors grew in number and took over the village.

Silmiougou, which originally had only a few Peul families living in huts and eking out an existence, today has around 1900 inhabitants, divided among 12 neighbourhoods. An initial consequence is that water is becoming scarce. People have started to supply themselves from a dammed-up pond fed by rain runoff. But during the dry season, the reservoir holds not a drop of water. Every woman who comes for water has to carry a basket. She uses this to remove the earth that she puts beside the reservoir to form a dike. The reservoir holds more water, but not all year. The village thus has to dig wells, to which have been added, over time, five boreholes hastily drilled by 'developers' without any involvement or participation of the communities. 'All we saw was some big machines that had come from Ouagadougou. They drilled some holes, and the kids went out to watch. Finally, people came to the chief of the village to have him tell us to drink that water', recalls one of the village inhabitants. Moreover, the boreholes are only a stopgap measure; some time later, two of them were rendered virtually unusable by polluted water.

The well: A battleground for women?

In September 1999, our research team visited Silmiougou as part of the GUCRE project. Consisting of university graduates (a hydraulic engineer, a sociologist and an economist) and a facilitator, our team looked into conflicts over water sources, particularly the way in which communities can participate in finding solutions.

In Silmiougou, the areas surrounding the wells and boreholes amount to boxing rings where women have a go at one another with unbelievable ferocity. Every day is replete with stories of broken water jugs, wounds of varying severity and family tensions. Cécile, the energetic president of the village women's association, describes the situation:

Every two or three days, the pump would break down. The men would search high and low for parts to repair it. They would accuse the women of causing the breakdowns through their daily disputes, and the women would point the finger at each other. That's when it would come to blows.

Indeed, it was the law of the jungle that prevailed around the wells. Everyone recalls the slogan of the time: 'Smash the jugs!' Old Cécile recalls those moments with a good deal of emotion:

After the dispute, the woman at fault would usually go home and tell her husband her version of the events. He would then come back to the well, armed with a machete. After the explanations, he would see that his wife was in the wrong. He would go home really angry and take it out on his wife. Then the wife would come back to the well and ask who the liar was that told her husband stories and made him scold her. And the fight would start all over again.

As if that were not enough, the children accompanying their mothers to the well would relieve themselves there and no one would pick up the excrement. The same thing happened with droppings from animals drinking at the well.

Tensions of many kinds

All over Nakambé, which is one of the four watershed basins of Burkina where the GUCRE team has been active, water is the source of conflicts. Those conflicts are exacerbated by power struggles. This is particularly true in Silmiougou, where the researchers set out in particular to study the local communication system. And what did they find? After the 1983 revolution, which brought Captain Sankara to power, the distribution of official government information was left to the village administrator, who represented the interface between the prefect and the local population. This new centre of power created tension within the village. The administrator and the village chief no longer even speak to each other. Two camps are forming, and their confrontations paralyse any initiative for the village's development. Elsewhere, as in Kora, a village located some 100km from Silmiougou, the situation is hardly better. Former residents of the village now living in the city quarrel with each other incessantly over anything to do with sponsoring development activities back home, to the point that any new initiative that the opposing camp might put forward is sure to fail. Assessing the overall situation, GUCRE wrote in one of its reports:

There are many conflicts: there are too few water points and they are badly managed; there is a shortage of drinking water; the pumps have broken down; there is no hygiene; people fight with each other at the wells over their social status, their religion or their ethnic background.

Tensions emerge whenever divergent interests, whether they are economic, political, social or cultural, meet around a water point without any civil, legal, administrative or customary mechanism for finding a compromise, a peaceful solution.

Facilitating social dialogue

This civil mechanism of which the researchers speak will involve instituting a culture of dialogue, an initial manifestation of which appeared in 2001 with the holding of a roundtable. This event brought together donors, government ministries (in particular, the ministries of the environment and of water, and their provincial offices), people involved in water projects and the local populations. The objective was to allow communities to present to policy-makers the results of their thinking in the wake of field surveys, and at the same time to learn about existing water management policies. During the roundtable, the experts pointed to a paradox: two-thirds of the well pumps are allowed to remain out of service, while communities face a drinking water shortage.

After the 2001 roundtable came the feedback phase, in which the conclusions were put to the communities and appropriate local solutions were sought. The team held discussions with local people about the causes of the pump breakdowns and possible solutions. It encouraged dialogue, joint efforts and interchange among stakeholders – communities, the research team, non-governmental organization (NGO) partners, project personnel and resource people – so that they could all help in solving conflicts relating to water use. To this end, the team resorted to various communication tools, such as discussion groups, a forum theatre and posters.

Local people took an active role in preparing and validating the messages during the discussion sessions. These meetings involved user groups that focused on issues of sharp concern to them, such as the cost of drinking water, the dysfunctional water management committees and the attitude of communities to the facilities. 'At the time', recalls Pascal Tandamba, the GUCRE facilitator, 'people did not feel particularly concerned about well management problems on the grounds that the wells belonged not to them, but to the government.'

The discussion sessions served not only to analyse the problems thoroughly and to review both local and imported solutions, but to select the most appropriate communication tools for resolving conflicts over water use. Thus, participants decided to produce a forum-theatre play with the village troupe based in Ziniaré, some 20km away. The content of the play reflected all the communication problems in the six villages involved in the project – namely, disputes over the well among women and between women and other users such as herders, who show up at peak hours to water their animals; hygiene; faulty water management (non-payment); and lack of infrastructure maintenance. Wherever the play was performed, the audience recognized themselves in the characters and in the plot. They were eager to join in the debate, which became very lively. After the performance in Bagré in eastern Burkina, another zone of GUCRE activity, a nun improvised a song that was soon on everyone's lips: 'GUCRE has come to untie our tongues. Let these practices go. Cohesion is the Royal Road to resolving our problems.' Apart from the theatre, communities also chose to use radio broadcasts, videos and posters. But a tool is not used for its own sake: it must meet several criteria, one of which is that it must be relevant to the objective at hand. It must also be properly adapted to the people who are going to use it. For example, the illustrations carried on the posters were drawn by the villagers themselves before they were redone by a commercial artist.

There are many accomplishments to point to: wells drilled or rehabilitated; water and health management committees established or reconstituted. The communities and the research team have drawn some lessons from all this.

The community viewpoint

People are pleased with the project team's approach, which is based on direct discussion with them and on developing an understanding of their own ways of communicating, and the nature of their social relationships. Every decision emerging from such an approach takes on greater credibility in the community's eyes.

Using discussion groups and pooling their results not only makes it possible to take account of the trepidation that women feel when speaking in front of men, but also facilitates the future sharing of ideas without gender distinction. People also feel that the project is not setting itself up in judgement over them, to award them good or bad scores. These are some of the conditions that facilitate the adoption of new practices, such as proper well maintenance.

The women are in favour of using this approach to decision-making, whatever the problem at hand: 'We could have a discussion like this about the mill to see how to manage it better and keep it running properly.'

Today, the issues under debate are rallying the entire community without any clan divisions: 'Amongst ourselves we often speak of Mr Pascal and of our two children, Judith and Alidobi,[2] and of the open spirit that he has brought us.'

The research team's viewpoint

The research team found that stepping up the pace of meetings among groups with divergent interests has helped to restore and strengthen social cohesion. The communities have begun to take a new view of the wells. People now feel themselves the owners of the wells and are therefore responsible for them. Disputes over the wells have declined considerably.

Social cohesion, then, constitutes the cornerstone for any process of participatory development, hence the importance of *Ned la to* ('man defines himself through his neighbour'). Nothing can be done without give and take, and give and take implies at least two individuals.

News of the project's success has been spreading. A number of villages along the river that participated in the project (for example, Pighin, Gweroundé, Ramitinga, Donsin and Voaga) have sent representatives to participate in the work as observers. The approach is winning converts even beyond the project zone, where people are calling upon the expertise of villages involved in the project to help them develop the same approach to resolving conflicts over water use, and in other matters as well.

Lessons learned

The experiment provided a number of lessons on how to improve the process. It is important to ensure that local people participate in defining the problem because there may be some differences in the way that researchers or facilitators see the problem and the way that the communities look at it. Researchers must not try to impose their own views of things or favour certain solutions over others that the community would like to try out. In other words, community participation must be effective right from the diagnostic stage and it must allow home-grown solutions to be sought, based on local experience and know-how.

In addition, it is important to create partnerships at all social levels, from the most important personality to the most marginalized. In effect, the most influential person may not necessarily possess the truth. We must also forge partnerships with individuals who, from close at hand or from far away, can influence activities undertaken with the local people. Those individuals must, of course, be identified in advance.

Problems and challenges

When experimenting with a new methodology, difficulties will frequently crop up, and it will be a challenge to address them.

Moreover, when conducting participatory development communication projects, we frequently find that people express tangible needs, such as women's credit, drilling a well or building a health centre. This raises an ethical question, for it will often be impossible to respond to all these material needs. The team is frequently asked how to deal with the situation, even though it has no immediate solutions to offer.

Another problem that the team faced was how to extend the impact of its intervention on a broader scale. No instrument had been established for capitalizing on the accomplishments or 'banking' them; therefore, it was hard to reproduce them in other villages of the Nakambé Basin or in other watersheds experiencing the same problem.

Finally, there is the question of the time and financing required to bring the process to a successful conclusion. Involving communities step by step throughout a participatory process takes a great deal of patience on the part of both the communities and of the other stakeholders, and it also requires a good deal of money. This point is not always fully appreciated. By means of this experiment, we can see that for a participatory communication endeavour to be effective requires the investment of enormous quantities of time and sufficient resources, in particular because of the frequent travel involved.

In conclusion, we may say that when communities participate in defining the problems holding back their development and in seeking solutions, they assert ownership over the outcomes, the benefits and the lessons. They will use this knowledge and these new ways of doing things to deal with development problems as they arise. This is the essence of the lessons that we can draw from the initiative, which used participatory communication to resolve conflicts over water resources in the Nakambé Basin.

Notes

1   GUCRE stands for Gestion des Usages Conflictuels des Ressources en Eau, or Management of Conflicting Uses of Water Resources, a project sponsored by the Centre d'Études pour le Développement Économique et Social (CEDRES) of the University of Ouagadougou in Burkina Faso.

2   In its work with rural communities, GUCRE collaborates with locally based facilitators. Judith and Alidobi were among these.

Experimenting with Participatory Development Communication in West Africa

Fatoumata Sow and Awa Adjibade

Long a neglected art, communication is today rightly treated as a key element in community development activities. Over the last decade, West Africa has seen numerous experiments for using the potential that communication offers people in finding solutions to their local problems. Many of these initiatives have sought to spark community dialogue about natural resource management. This chapter looks at two such projects, one for managing water use conflicts and the other focusing on rural communication and sustainable development.

Although the authors were not directly involved in these projects, they were associated with them in several ways, and from their observations they have drawn some lessons that could help to improve existing practices in participatory development communication (PDC).

Today, ten years after the first experiments with participatory communication got under way in West Africa, we can say that far from being cut off from local reality, this approach finds its wellspring in community life, in inter-group relations, in local knowledge and know-how, and in development practices and activities, as well as in existing systems and tools of communication.

In these communities, participatory communication has helped to identify the major concerns and conflicts related to natural resources. Disputes typically have to do with land occupancy or with crops trampled by livestock; disagreement over the location of a well or the boundaries of a property; or tensions sparked by deforestation through bushfires and illegal woodcutting. Power struggles between customary leaders and government authorities over the question of landownership also give rise to many problems.

These conflicts and tensions crop up so frequently that they can shatter the harmony of village life and pose an obstacle to local development. The lack of consensus about the underlying causes and on mechanisms for managing conflicts within the community or between neighbouring villages, and the lack of modern legislation governing the notion of property itself are just some of the elements that require good communication if local development is to be sustainable. The PDC approach then becomes a useful tool in identifying the sources of conflicts, and in seeking and implementing solutions and sustaining efforts for the sound management of natural resources.

By promoting dialogue and the horizontal exchange of views, PDC allows all social groups to contribute to resolving problems. With this approach, religious and customary chiefs, government experts and authorities, youth, the elderly, men and women can create the dynamics for sharing experiences and forging partnerships without distinction as to social class or caste, dynamics that can work not only within the community but between several communities or villages.

Another important outcome has been to raise the status of women, evident now in their readiness to speak in public, in the emergence of women's groups and in the greater openness of mind that characterizes women and men alike. Thanks to the knowledge and expertise acquired through these initiatives, women who were once excluded from outreach sessions have now themselves become agents of training and awareness raising for the community (and even for their husbands). They are quicker to speak out, they are putting on theatrical plays and they are creating new channels of access to credit, training and HIV/AIDS prevention.

A greater appreciation of local know-how is a further contribution that participatory communication is making to development. By enlisting people in diagnosing problems and seeking solutions, the project teams have offered them the opportunity to share their ancestral know-how and practices. This has helped to mobilize and involve local people who are no longer expected simply to accept and apply the advice they receive from researchers or technicians.

Conditions for success

If participatory communication is to be really successful, there are a number of conditions that should be met. Through their physical presence at the side of local people and participating partners, facilitators also play a key role in participatory communication while serving as relay points for field coordination. Their commitment and their mastery of the approach are important for achieving the expected results.

The process can be kept on track through an iterative approach that includes regular follow-up activities involving dialogue, clarification, training and capitalizing information. In other words, if all the stakeholders can agree at the outset on a possible solution and on ways of implementing it, there is no reason for it to fail.

If all stakeholders have a proper common understanding of the concept of participatory communication, this will avoid frustration and resistance that could hold up activities in the field. Stakeholders can be brought up to speed by organizing workshops where key players and partners are introduced to the concept of PDC, discuss it and seek clarifications, and through meetings to coordinate follow-up and support in the field. Regular feedback on results will provide further encouragement to participating communities and can be used to help spread the word about the activities.

Applying the PDC approach to concrete local concerns can facilitate an understanding of the initial context in which the activities are being pursued and an appreciation of the objectives to be achieved. It also helps to rally communities to resolve their problems and to take responsibility for implementing the adopted solutions, something that is essential in asserting community ownership over projects. Of course, keeping both the customary and administrative authorities informed and interested will contribute greatly to achieving the objectives and must not be overlooked.

Interpersonal communication is also very important in the PDC approach. This means that the facilitators must first be given proper training in interpersonal communication techniques.

An iterative process for evaluating progress and a mechanism for hands-on monitoring and coaching will help players to ask the right questions as they go along and make the necessary corrections, and thus take a qualitative leap forward in achieving project objectives.

Project governance can be enhanced by setting up a management committee composed of and led by community members. This will not only build mutual trust and maintain harmony within the project, but will also bring it transparency, which is essential if community members are to rally to the process and remain committed to it.

Participatory communication facilitators: Aptitudes, qualities and roles

Experience has shown that for the PDC approach to succeed in natural resource management or in any other development project, the facilitator or the person in charge of the process must be:

- attentive and willing to listen to others;
- available to pitch in at any time;
- considerate and respectful of other partners;
- sufficiently imbued with the subject and the problems to be addressed;
- able to put forth persuasive ideas;
- considerate of the local setting;
- open and readily reachable;
- motivated, confident and devoted to the task; and
- recognized and admired by the community and the authorities.

The facilitator, as the fulcrum in the PDC process, has the essential role of:

- reducing tensions (resolving social and cultural obstacles);
- putting specialists in contact with participants;
- fostering a good working relationship among all partners;
- engaging in advocacy work with donors and development partners;
- identifying the specific needs of partners and different local groups;
- coordinating all activities;
- negotiating agreements between groups, partners, authorities and donors;
- bringing success stories to the attention of all partners and local participants.

Problems encountered in the participatory communication approach

As a process for achieving social change and resolving communities' development problems, PDC is a long-term undertaking and a number of stumbling blocks may be encountered along the way.

The participatory communication approach can also run into other problems that communication alone cannot resolve. Moreover, the approach may well take longer than the planned life of a project and this may not fit with the time horizons of donors, partners or local people.

Budgetary constraints may affect the frequency and duration of the follow-up that such activities require. There are also bits of social and cultural baggage that make it hard to change mentalities and behaviour, something that takes a lot of time. Finally, managing power relationships within communities demands a great deal of skill and finesse.

Conclusions

The approach championed by participatory development communication can achieve results of great significance for communities, researchers and technical partners by fostering initiatives that involve dialogue about managing environmental problems. Experiments in West Africa over the last decade have provided clear indications that this methodology can be applied very widely to natural resource management projects and to any other development project. Yet, the facilitator must be careful not to arouse false expectations. Nor should the facilitator try to take the place of development agencies. Finally, the facilitator must refrain from imposing preconceived solutions and must try, instead, to promote the kind of synergy in which the community itself will take charge of its development.

The situation in Africa today, marked by poverty and recurrent environmental disasters, calls for a real break with the past in terms of community participation and the place and role of communication and information as key factors in reinforcing people's capacities for self-development.

The boom in community or co-operative-run radio stations in the countryside, and the steady spread of information and communication technologies are obviously important factors that must be taken into account in planning strategies for strengthening and expanding participatory development communication.

Strategic Communication in Community-Based Fisheries and Forestry: A Case from Cambodia

Jakob S. Thompson

Faced with tremendous communication needs, many natural resource management projects undertake to develop materials and implement activities without a spelled-out and systematic use of communication. When the need for a clearer and more effective strategy arises, reflecting on past and future challenges can help to select the most appropriate methodology to respond to the needs of communities.

This chapter looks at the numerous difficulties faced when dealing with a very large array of participants and communities where cultural, occupational and geographical features seem to divide people rather than unite them. It is based on the reflection of a training and extension officer whose appreciation of development communication has been evolving along the way in view of the multiple challenges faced by a community-based fishery and forestry initiative in Cambodia. As a result, the approach has evolved from the production of materials to the systematic use of communication tools and, more recently, to the establishment of more participatory processes.

The Cambodian province of Siem Reap is located in the north end of the Tonle Sap Lake, which is the largest freshwater lake in South-East Asia. Every year, the flow of the river connecting the Tonle Sap Lake to the Mekong River is reversed because the flood level in the Mekong becomes higher than that of the Lake. This results in a fivefold increase of the area covered by the lake. During that time, the productive inundated areas become the nesting, feeding and spawning ground for hundreds of different fish, bird and animal species. Thus, this seasonal flooding is a condition for the existence of this rich ecosystem, which provides a source of food and income for a rapidly increasing population on and around the lake.

The overall development objective for the initiative, entitled Participatory Natural Resource Management in the Tonle Sap Region, now in its third phase, is to establish responsible, productive and sustainable management of forest and fishery resources by local communities in order to meet local needs and stimulate local development within the province of Siem Reap. More specifically, it aims at developing community fisheries and forestry, while promoting private and community-based development activities in support of natural resource management. It also pursues the objective of strengthening institutions and building local and regional capacity.

The initiative emphasizes the facilitation of productive and sustainable community-based natural resource management (CBNRM). In the season-ally flooded zone around the great lake, the initiative works with fishing com-munities to protect and manage thousands of hectares of inundated forest habitat for the production of both fish and forest products. In the upland areas, the concept of community forest management is spreading as communities seek to protect and conserve rapidly disappearing forest resources. Currently, the initiative is providing support to the provincial government departments of forestry, fishery and environment, which assist 44 community forestry organizations in the management of more than 20,000ha of forest lands and ten community fishery organizations in the management of close to 108,000ha of fishing grounds.

Besides community-based natural resource management, the initiative has identified and engaged in a number of supporting activities. Tree nursery development and the distribution of seedlings to villages and farmers, along with aquaculture development, has been introduced as a livelihood option to reduce people's dependency upon the natural resources base. Rural credit and training of savings groups is stimulating local development, and environmental education for children and adults has become a priority issue, with a view to strengthening the environmental awareness of current and future natural resource managers in Siem Reap. A geographical information system (GIS) unit is also providing maps and mapping support to the activities.

The communication challenge

The community fishery and forestry processes are participatory step-by-step approaches that start with a range of assessments. The first steps aim at setting up a management committee through village- and community-level elections; drafting, finalizing and gaining approval for bylaws and statutes; demarcating the community forestry or fishing grounds; and supporting the development, approval and implementation of a management plan regarding the use and protection of natural resources by the community in years to come. Open and effective communication in a broad sense is a key agent in making this community fishery or forestry process work: communication between project staff and stakeholders, between stakeholders and also among staff members. Communication also plays an important role in influencing and strengthening the external framework for community-based natural resource management, as well as in gaining the necessary recognition and support from authorities, decision-makers and other key actors.

As is the case in most countries, there is not a strong tradition for inter-departmental cooperation in the Cambodian governmental structures. The multidisciplinary approach of the initiative has, however, depended upon and created an opportunity, as well as a climate, for cooperation and exchange between the forestry and fishery departments, and with the Ministry of Environment. However, the existing power structures in Cambodian society are complex. In some cases, the lack of transparency and control functions makes it difficult for open and unambiguous communication to take place.

Variety of communication challenges

For a number of reasons, the communities involved in CBNRM in Cambodia represent a wide range of communication challenges. At one end of the spectrum, the initiative works to assist community forestry development in sites with one or a small number of villages located along a small road or within a limited geographic area. The people involved seem to have a sense of community belonging and they come from similar ethnic and religious backgrounds. To a large extent, people in these communities have corresponding priorities when it comes to the use and protection of forest resources. In this setting, a communication or extension worker can achieve a lot by arranging a village meeting with the help of the village chief or by posting information on the village information board. The success of a meeting or an information campaign in such a setting largely depends upon the quality of its planning and implementation, and not upon factors such as logistics, community structures or differences in interest.

At the other end of the spectrum, one can find community fishery sites on the lake, with multiple villages of different sizes and structures. Their inhabitants represent different ethnic groups with a variety of languages and religions. There is a wide range of resource users, from the fishermen who entirely rely on fishing for their income and livelihood to the rice farmers who see cleared agricultural land as more valuable than inundated forest. In addition, there is also a wide range of people influencing the priorities of fishermen: fish buyers and moneylenders have stakes in their catches; government officials influence the way in which they fish through the exercise of their power; offenders overexploit the resources with no long-term responsibility and often use fishing methods that are damaging to the environment; and migratory fishermen/farmers from outside the community settle by the lake during the best fishing season. Furthermore, tourism is becoming an increasingly important source of income for some fishing communities close to Siem Reap.

Besides the complex human aspects of a fishing community, the physical features also represent a challenge to the communication facilitator: a floating fishing village is in a constant state of moving. People relocate their floating houses about 15 times per year in order to keep up with the changing water level of the lake. After the rainy season, most houses are moored along a road with the morning market right on the doorstep and the main town of Siem Reap or another main market within easy reach. Later, in the dry season, the water level of the lake has dropped by 7m to 8m and the same houses are now located many kilometres further out, on the open lake. At that time of year, a trip to town involves a boat trip and a long and painful motorbike trip through the inundated zone, only to get you to where your house used to be some months before. When entering the inundated forest on foot at this time of year, one also has to look up to see the information signs about community fishery that were previously posted in the treetops during the wet season. In this setting, there are many obstacles to effective communication, and the skills and intentions of the communication or extension worker are sometimes not enough to inform and engage the people in the community. The communication process therefore has to rely on a much wider range of communication tools, and one has to build more effectively on local capacities in order to reach an audience or to encourage broad participation.

The communication approach

Preliminary assessments and experiences indicate that regardless of the initiatives under way, people rely on direct interpersonal communication for their information needs in the sense that people talk to each other to convey information. People see radio and meetings as the potentially best way of communicating within their communities; but so far, the mass media has mostly been used for entertainment and, to some extent, for news. Written materials or newspapers have little potential due to very low levels of literacy. Illustrated posters have a potential; but the lack of permanent structures and the harsh climatic conditions pose a challenge as to when they are to be produced and posted. Illustrated handouts or leaflets have been identified and used by the project as a means of communicating; but again, only limited information can be transferred in the form of pictures. However, when text is used, the understanding is generally quite low.

The initiative has used a number of different tools and approaches to communicate to, from and within communities and stakeholder groups in order to build awareness, inform, educate and encourage participation. These tools have primarily been developed by the initiative and participation has so far been a result of communication activities, rather than being used for building the communication itself. However, attempts have been made to include communities in the development of communication tools. Nevertheless, so far their participation has been limited to consultations on the content and layout of single tools, rather than full-fledged strategy development with participatory rural communication appraisal (PRCA) or similar tools.

Meetings, workshops, etc.

In this initiative, meetings are the main communication tool. At the community level, the management committees, assisted by the field facilitators, invite people to attend meetings to present and discuss different aspects of the community forestry and fishery processes. The drafting of rules, regulations and management plans depends upon the participation of community members and other stakeholders, and fair elections obviously depend upon all the groups participating. Issues of such importance that plenary discussions are required also arise independently of the community fishery or forestry process. Meetings are then called in order to inform and consult the views of the people. The project is also facilitating horizontal communication between communities in so-called community fishery and forestry network meetings, to which representatives of all the communities involved in CBNRM development are invited. These meetings give the communities a chance to communicate among themselves and to communicate directly with representatives of the provincial and centralized authorities. The monthly district-level planning meetings have also been another important channel of communication. At earlier stages, the project team leader would personally attend the meetings; but in the past years, this responsibility has been handed to field staff. The main goal is to inform community representatives about community-based natural resource management and to gain recognition amongst stakeholders. Requests for further expansion of community forestry to new areas are commonly put forward by communities in these meetings.

Extension and environmental education for adults

On the lake, assessments indicate that meetings are too time-consuming and difficult to attend for the working population. In order to raise general environmental awareness in a wider audience than the meetings normally attract, the environmental educators travel from village to village with an environmental educational programme that is conducted at a time of day when fishermen can attend (this time varies depending upon the season and the type of fishing gear they use). A video about the history and current state of the natural resources in their community is used as an entry point to a problem tree exercise that eventually leads to a discussion of the possible solutions to the problems they face. The environmental education is based on entertaining and highly interactive approaches, small group discussions and plenty of room for humour and flexibility.

Besides destructive fishing, the encroachment of forest for agriculture and the extensive use of pesticides, which is likely to affect the long-term fish productivity of the lake, also constitute environmental concerns. Pesticides are likely to influence the safety of eating fish from the lake. As a result, the project has used a farmer field school approach with the seasonal farmers who grow agricultural crops on a large scale within the inundation zone. Evaluation indicates that these initiatives have led to an increase in awareness, improved farming practices and a reduction in pesticide use.

There is also the case of migratory people who enter the community fishery areas to fish every year. Community fishery organizations are assisted in their work of communicating with these people. Typically, these people are farmers who spend the dry season in temporary shelters close to channels or ponds in the community fishery areas. They fish and process fish for a few months before going back to celebrate the Khmer New Year and resuming agriculture, when the rains eventually start again. This is considered a traditional practice that may prove difficult to curb. Consequently, there is a need for information about the local rules that apply in the community; fishing fees; the basic ecological reasons for banning some fishing gear; and about any management practices in the community that they could take part in, in order to contribute to the sustainable management of the resources. To achieve this, community members and project facilitators travel around the inundation zone by boat and on foot to localize and inform people as they go.

Environmental education for children

Cooperation with the provincial Ministry of Environment has primarily evolved around an extensive environmental education programme. Essentially, the programme includes the establishment and use of a floating environmental education centre on the lake called the Gecko Centre and the development and implementation of an environmental education manual for primary school teachers with three partner non-governmental organizations (NGOs).

The Gecko Centre is located in the floating community closest to the main town of Siem Reap. Twice a week, the Gecko and its staff host a day of environmental education activities for groups of schoolchildren in the area in order to raise their environmental awareness and knowledge about the lake ecosystem in particular. The Gecko Centre also works as an entry point for some of the tourist boats that are taking people on excursions to see the natural and cultural environment on the lake. Besides contributing to the general awareness about the lake, this is providing an opportunity for the Gecko to cover some of the operating expenses from voluntary contributions made by tourists or other visitors.

The environmental education programme also includes the design, production and use of an environmental education manual in Khmer. In collaboration with three other NGOs, a manual for teachers has been developed and 1000 copies have been distributed for use in primary schools. Some 100 teachers have already been trained in environmental education principles and how to use the manual. They are now implementing the book as part of their work. Today, monks are also being trained in how to implement environmental education activities in the pagodas for out-of-school children. The book will be updated and reprinted based on the results of this first phase in order to augment environmental education methods and the environmental curriculum in the Cambodian school system.

Supporting awareness building

Posters are developed and/or used by the project with general or site- or topic-specific content, and the community fishery/forestry organizations have also been supported financially to build signboards for information dissemination, general environmental awareness raising, etc. The posters are hung and the sign boards are fixed at central points in the communities or at places that fishermen visit, such as landing sites for fish or outside the pagoda.

Because it is located close to Angkor Wat, Siem Reap is a major tourist destination. Therefore, it was deemed important to contribute to the general awareness raising by disseminating information on the importance and unique features of the Tonle Sap Lake ecosystem and community-based natural resource management. In addition to the Gecko Centre, visitors will find posters, books, visiting cards and locally produced palm sugar candies on display or for sale at the airport, as well as in many hotels and restaurants in town. At the provincial Ministry of Agriculture, there is a documentation centre with project documentation and other relevant documents, books, videos and resources. This centre is being used by project staff for information searches; but it is also open to and frequently used by students, NGOs and other users. A richly illustrated coffee-table book with written contributions from local and international experts on the culture and nature of the lake is the latest contribution to increasing the international audience's awareness of the importance of the Tonle Sap Lake.

Communicating management activities

Another challenge for the community fishery and forestry organizations has been to make everyone aware of the existence of CBNRM and to build acceptance for the activities and the rights and responsibilities that community-based natural resource management implies for the intended members. This is being done through the registration of members and the collection of a small membership fee. Since people usually do not give money for something they know little about, they start asking questions about community fishery and forestry.

Other practical management activities, such as the development of fish sanctuaries, have also proved to engage people and to gain general acceptance for the role of community fishery. The large fenced-off areas are built to shelter a stock of brood fish throughout the most intensive fishing season. The fish can then migrate to the forest in the flood season to spawn. Coloured flags and explanatory signs on each corner mark where the sanctuaries are, and people don't pass one by without asking what it is for. In community forestry, the strips of cleared forest and vegetation to prevent fire from spreading have much the same effect, along with thinning activities and marked research plots.

Why a strategic approach to communication?

As this overview suggests, a wide range of tools and methodologies are applied in order to inform and educate people, as well as to promote participation. However, there has not been a clear strategy for communication work, and the different activities have not been applied as part of a larger communication plan. The general approach has, instead, been to apply available and appropriate tools to deal with single communication issues as they arise. As mentioned earlier, the increasing communication needs of the project have resulted in the establishment of a training, extension and communication unit that is supposed to deal with communication needs in a more systematic way, and in accordance with communication for development principles. This was also a response to a possible expansion of project activities to other provinces in Cambodia. In hindsight, one could argue that the project probably would benefit from a more systematic approach to communication for a number of reasons, as follows.

Improving effectiveness

Throughout the years, it has become clear that communication materials that do not involve people in the development or interpretation process have very limited impact. Consequently, the project involves community members in, for example, making specifically designed community fishery posters for each site. The management committees in the communities and the field facilitators seem to have great expectations regarding the effectiveness of such materials. However, the use of single materials outside a well-planned and properly scaled communication strategy seemingly works against its purpose. First, the huge needs of the communities to disseminate information cannot be covered by a single material such as a poster without seriously reducing its quality and effectiveness, even after a strict prioritization exercise is done with the community to limit the amount of information squeezed onto the poster. Second, the isolated use of a single tool such as a poster may be used to transfer the committees' responsibility of informing people to becoming informed, since the information is now theoretically accessible to the people in the community. The management committee therefore seems to think that a poster relieves them of the important task of talking with people and exploring other complementary means of communication in order to ensure that everyone has a chance to express their view.

Making better use of multidisciplinary aspects

The initiative strives to apply a multidisciplinary approach to deal with complex real-life issues. The project, however, works within governmental departments, and the development and application of cross-sectoral solutions in a sectoral system is one of the project's main achievements. However, an overarching communication strategy could probably enable the project to benefit even more from the multidisciplinary approach by assessing strengths and weaknesses in the different sectors as a basis for exchanging communication expertise and tools between sectors.

Building stronger project and community identity

Tonle Sap Lake is increasingly becoming the focus of small and large development initiatives. This has resulted in a multiplicity of actors with complementing, overlapping or even contradictory objectives and methods compared to those of the initiative. In this setting, it would be beneficial to clearly point out to people what is, and isn't, community-based natural resource management development. In the future, communities will probably have to make the distinction in order to avoid confusion and to decide what is worth spending time and effort on. If not, they risk spending too much time accommodating researchers, NGOs or other actors, with limited potential benefit for the development of the community or even with agendas that do not benefit the communities at all.

Learning lessons

Based on our impression and in-house assessments, the work to raise awareness, inform, educate and communicate with communities has clearly had an impact within the communities. People generally know that community fishery and forestry exist and their purposes. People are also ready to engage in the management of natural resources and the environment since they generally understand the importance of the resource base and their effect on it. It is, however, difficult to assign the successes to any part of the approach, or even to decide if the improvement is a result of this initiative or of other factors, since the lack of an overall strategy also means that there is no established system for monitoring and evaluating the communication. Of course, this also makes it more complicated to develop guidelines and recommendations for the future, or for others to learn from, based on the initiative's communication approach.

The initiative has worked closely with partners and within governmental structures to lobby and advocate for the importance of community-based natural resource management. It has also contributed to the formulation of policies on CBNRM. Some of this work has been obvious in speeches and written contributions and comments; but much of it is also achieved through less visible channels and media. This aspect of the initiative's impact might be one that is strengthened due to the somewhat random but, nonetheless, flexible approach to communication in the project. Again, the documentation, lessons learning and reproducibility of this aspect of the project will be difficult to ensure since its success is built, to a large extent, on personal devotion and communication skills. Thus, it may be difficult to make generalized recommendations or to propose specific methodologies.

Scope for participatory development communication

So far, there has been a strong demand in Siem Reap for expansion and for handing over new land to communities who will manage it. The framework for development has, by and large, been set by factors outside the communities, such as the formal requirements for community-based natural resource management. The further development and continuation of CBNRM activities, also beyond the lifetime of the initiative, will need to look at diversifying community-based natural resource management activities to accommodate a wider range of people in the communities. Strengthening CBNRM and achieving widened support will therefore depend more and more upon the participation of the people who benefit from, and take part in, the management of the resources. In other words, there is a need to allow people to interact, collaboratively learn and influence decision-making in order to make CBNRM suit their needs. Participatory development communication approaches can potentially identify the community interventions and management options that are best suited for each community in order to ensure the basic CBNRM ideals of participation, equity and sustainability. The sometimes complex nature of the communities will probably also call for strong communication efforts in order to fulfil the intentions of any intervention. In other words, PDC is meant to play an important role both as a tool and as an outcome of CBNRM in Siem Reap. However, in order for this to succeed, some important preconditions must be met:

- Skilled CBNRM facilitators with an understanding of participatory processes and an appreciation of communication as a tool for development must be supported in working with communities for the benefit of the latter.
- At all levels, there must be unconditional support for the communities' right to take the lead in managing, identifying and implementing management-related activities that sustain CBNRM.

Paving the Way for Creating Space in Local Forest Management in the Philippines

Cleofe S. Torres

When, after decades of dispossession and disempowerment, shifting government policies make it possible for a devolution process to take place, local people's organizations can only rejoice. But assuming full responsibility for the environment they live in requires that communities acquire new skills and new capacities, while building their social capital. Although not a panacea, participatory development communication (PDC) can greatly help communities to take up the new challenges associated with decision-making and self-determination.

The Bayagong Association for Community Development Inc (BACDI) is an upland people's organization in Aritao, Nueva Vizcaya, the Philippines. Its members were victims of the involuntary relocation project brought about by the government's construction of a dam that submerged the people's ancestral lands. Instead of taking the government's offer of resettlement in a place not akin to their culture, they migrated to a nearby forested land. This experience precipitated their engagement in some kind of participatory development communication as they needed, then, to discuss their fate and how they could cope with it. Of course, these people did not know that they were involved in PDC!

Labelled as squatters and encroachers by government, they struggled for almost 28 years to hold on to the lands that they have been *de facto* occupying since they migrated in 1960. Periodically during the past, they had to resist and endure the efforts of government agents and threats from outside speculators to expel them from the area. To make their appeals known to the concerned authorities, these upland farmers engaged in dialogues with government officials and authorities. But when their voice seemed not to be heard, they resorted to street protests and lobbying. These strategies were outcomes of their frequent discussions and analysis of their situation.

A wind of change worked in BACDI's favour when the government's policy on natural resource management adopted community-based forest management as the national strategy for sustainable development of open forestlands. BACDI applied for and was awarded the Community Forest Stewardship Agreement, a tenurial instrument that formally legitimized their claim over the forestland they have been occupying. It granted them tenure of 25 years, renewable for another 25 years.

While the political and social intent of such a devolution strategy was highly appreciated, issues emerged regarding the community's readiness and absorptive capacity to handle the devolved tasks and functions. Devolution as used here refers to 'the process where the locus of power and control shifts from the state to the local communities' (Magno, 2003). Under community-based forest management and as part of their new set of responsibilities, BACDI members underwent the process of participatory resource management planning. Here, PDC took the form of a social preparation method, equipping the people with the knowledge and skills that they would need as forest resource managers. Through community mapping, resource identification and problem prioritization, they learned to draft their plan and called it their own.

The other gender-sensitive participatory rapid appraisal methodologies used for situational analysis included community resource profiling; gender-disaggregated household activity and decision-making; stakeholder identification and analysis; political-ecological mapping; historical-structural analysis; institutional analysis; participatory analysis of problems and options; and resource management action planning. In all these methods, PDC was a core process for learning and planning.

Supplementing this variety of PDC methods were communication tools such as Venn diagrams, maps, community billboards and posters, written documents of their organization, and policies agreed upon for managing their resources.

This exercise paved the way for the community members to have a better grasp of the quality of their resources, their cultural integrity, their social capital and their political capacity, as well as a sense of their prejudices, weaknesses and vulnerabilities – knowledge that they could have not unravelled had they not undergone PDC. This also enabled them to identify internal and external threats to their resource management and how these might be handled.

To a large extent, the participatory experience enabled them to implement their plan with more confidence and better direction. Whereas before, the community simply remained buried in their culture of silence and subservience, the knowledge acquired of their rights and responsibilities through community meetings and discussions enabled them to become more open and assertive. As a result, when they conducted social mobilization among their members and allies, they were already more aware of community organizing, government policies, their internal capacities, external possibilities and the given constraints in the field.

Participatory development communication paved the way for BACDI's link-up with other government and non-governmental organizations (NGOs). Members knew that they could not muster all the resources needed to manage their forest resources well. They eventually realized that they also needed to access external assistance, such as livelihood opportunities, *barangay* roads, markets, credit, schools, health centres and other social services. Partnerships became another imperative, and they were able to establish alliances by using PDC.

As they learned more about themselves and their resources from the various participatory rural appraisal methods used in the study, BACDI members were able to come up with more rational plans and approaches for managing their communal forest. In the past, they had left these matters to their leader and whoever was deemed influential in their group. PDC also enabled them to reorient their outlook and address their needs by first using locally available resources before turning to outside sources for assistance.

For monitoring and evaluation, they employed informal, unstructured discussions where they would ask each other and reflect upon where they were at a certain time relative to their plans and targets. The observations they made were then discussed in their regular community meetings. Here, members were allowed to clarify and validate their observations.

Used in the various facets of participatory resource planning, PDC has actually served as a mechanism and, at the same time, as the context of group learning. As they progressed in learning about their biophysical, socio-cultural, economic and political environments, BACDI members became more enlightened and rational managers of their forest resource. Even though most of them had only attained a low level of education, and some were even unschooled, they felt that they had learned many things from their community natural resource management activities. As a result, they were able to delineate the bounds and limits of the physical and political space that they had claimed for themselves. This was reflected by the sample plan and set of policies formulated governing the sustainable use and protection of their resources.

Social dialogue became a frequent activity in which they engaged by virtue of the communal nature of their forest resource. Decisions had to be collective and inclusive. Although discussions were not always smooth sailing, the members of the community have gradually learned the techniques of negotiation and consensus-building. In the process, their dwindling social capital has been enhanced. Social capital refers to cooperative social relations and collective action processes. This is embedded in 'norms or reciprocity, networks of civic engagement, trust, and obligations that facilitate coordinated activities' (Asian Development Bank, 1994). Since social capital depreciates through time, constant dialogue and open communication prevented it from deteriorating. Strong social capital has enabled the BACDI community to protect their traditional resource system from outside encroachers and the centralizing tendency of the state.

As BACDI members learned to master the science and art of forest management, they were continuously bombarded with challenges. Dominant among these was the multi-stakeholder setting for decision-making and action planning. It was through PDC that community members were able to respect and manage the diversity of views on an issue. Constant dialogue opened them to many possibilities and made them realize that there could, indeed, be many possible solutions to a problem.

Likewise, PDC provided the venue for mainstreaming women in a men-dominated BACDI ethnic group. From mere cooks and servers during community meetings, women started assuming more substantive roles, such as being a liaison officer, secretary, treasurer and even *barangay* councillor.

It must be emphasized that it was not PDC alone that paved the way to local forest management by BACDI members. Other factors were also involved, such as forest culture, social capital, policy presence and assistance by external actors. It can be safely assumed, however, that participatory development communication played a critical role, tempering the socio-political environment, internal and external to the community, so that a climate favourable to the community's takeover of resource management was created.

Reflections

Just like any other communication methods, PDC is not a panacea. It also has certain caveats, nuances and limitations. As a tool, its users and practitioners have to understand its profound intricacies. It takes time and practice, and a lot of learning from failures, to appreciate what it can and cannot do. Hence, its participants have to understand the action–reflection–action dynamics that are built into it.

Participatory development communication, as a catalyst for change, should also be accompanied by the inputs necessary for development, such as credit or capital, roads, water and technical assistance. It can establish the link between and among the development players in the community. Information and consensus are necessary conditions; but they are not sufficient to bring about broad development as desired by the community.

Likewise, devolution does not necessarily ensure effective local forest management. Policy and political and technical support have to be provided. But the use of PDC as an ingredient in the process makes it more workable, socially acceptable and satisfying in meeting the democratization and efficiency agenda of natural resource management. Moreover, PDC enables the evolution of 'participation-as-engagement' process, veering away from the usual 'participation-as-involvement' process (Contreras, 2003). The latter simply implies that the 'subject' or actor is merely a participant in the process, in a position of powerlessness. The former, instead, transforms the actor into an 'active subject' and not just an object or a client. 'Participation as engagement' is a necessary condition for genuine empowerment and for creating space in local forest management, which can be best achieved through participatory development communication. The impact on the community, especially in terms of managing their own forest, would have been different had other forms of top-down communication been employed.

References

Asian Development Bank (1994) *Handbook for Incorporation of Social Dimensions in Projects*, Asian Development Bank, Manila, the Philippines

Contreras, A. P. (ed) (2003) *Creating Space for Local Forest Management*, La Salle Institute of Governance, Manila, the Philippines

Magno, F. A. (2003) 'Forest devolution and social capital', in Contreras, A. P. (ed) *Creating Space for Local Forest Management*, La Salle Institute of Governance, Manila, the Philippines

IV

Communication Tools and Participatory Approaches

Communication Tools in the Hands of Ugandan Farmers

Nora Naiboka Odoi[1]

Introduction

Participatory development communication (PDC) aims to establish two-way horizontal communication processes. However, in a situation where there is a need to introduce new information, communication processes can hardly be horizontal if the tools used to communicate with farmers remain in the hands of experts and professionals. This is what the research team of the National Banana Research Project in Uganda had in mind when it established a farmer-to-farmer training programme. Indeed, farmers were so empowered

by the process and enthusiastic about sharing their new knowledge with other farmers that they soon wanted to produce their own communication material. After pushing aside the material that had been initially produced for them, they assessed their communication needs, established objectives, defined activities, produced their own material and even selected indicators to measure their success. In other words, they took control of the communication process from beginning to end.

Farmer-to-farmer communication

Quiet reigned inside the sub-county headquarters building as farmers of Ddwaniro, within the south-western district of Rakai in Uganda, looked with apprehension at Nora and Moses, the research team's communication resource staff. Nora and Moses fidgeted with the television screen and video deck while Fred, the driver, fixed the power generator. As farmers peeked at the TV screen, they all seemed to be saying the same thing in their hearts: 'Let nothing go wrong now with this video. This is our chance to show other farmers what we have learned after all these months of learning and practising.'

The loud blast from the electric gadgets signalled that all was well and ready with the machines. Moses turned down the television volume as Nora began:

> *We can now settle down and watch what we recorded last time. Remember, this is the video we are going to use to share knowledge about proper banana management with other farmers. We are nearing the launching day of our farmer-to-farmer teaching programme when we shall invite other farmers, district leaders, researchers and all members of our community and neighbouring communities to show them the results of our endeavours.*

The farmers smiled at the thought of moving from the production stage onto the next stage when they would show other farmers what they had been doing during these past months.

The video began rolling. Farmers were visibly excited as they recognized themselves in the film:

> *That, indeed, is Mr Kubo talking. We saw him doing that... That is, indeed, his well-kept banana garden... But why is he talking from far away? It would be better if he were talking while standing near us [in the foreground]... Moreover, he is not looking at us from the screen... The pictures are not flowing well. We should not have used Mr Muganda to illustrate the mulching technology. We could have used Mrs Muganda instead... She is better at explaining issues. But where is the good banana bunch to show the result of a well-kept banana garden?*

The video continued to a stop just in time as numerous hands shot up simultaneously. Mr Kubo began the barrage of remarks that followed the viewing of the video clip: 'Madame, we have seen the video; but I for myself am not convinced that it will deliver our message.' The farmers unanimously rejected the video, agreeing with Mr Kubo and adding that they could produce a better video than that, but that this time they were going to make a more thorough plan of action. This marked the end of the farmers' meeting with Nora and Moses. The farmers immediately reconvened their own meeting and chose a chairman for the session. The agenda was how to produce a better video than the one they had just watched.

The farmers chose Mr Sebulime to be the overall presenter of their farmer-to-farmer teaching video. Mr Kubo was selected to demonstrate organic manure-making; Mr Lubwamira would demonstrate the digging of trenches in order to guard against soil erosion; and Mrs Muganda would explain the proper mulching of a banana garden. They fixed a date for the next video recording session and informed Nora and Moses about their decision.

On the day preceding the video recording, Nora and Moses arrived in Ddwaniro, together with a professional cameraman. They made contact with the farmers, who took them around the banana gardens that were going to be used for the video illustrations. A mini rehearsal was organized to determine what was going to take place the next day.

Farmers took the lead on recording day. They guided the team of researchers and the cameraman to the different banana gardens that were going to be used for demonstration purposes. They had prepared so well that they knew exactly what sequence the recording would follow. Consequently, the recording process took only a short time, and there was little footage to cut out during the editing of the video. Although this last phase took place in Kampala, away from the farmers, they unanimously accepted the video the next time that it was shown to them. They agreed that other farmers would understand and probably make use of the message contained in the video. This was confirmed later when they showed the video to other farmers of Ddwaniro.

Using photography

Slowly shaking his head with a sneer, Mr Sebulime examined the photographs that Moses had brought back from Kampala.

'These pictures cannot do. Look, the women appear as if they are going to a wedding feast. How can a farmer working in a garden dress up smartly like this?'

'That is exactly what was recorded last time', Moses replied.

'But it is not right,' Mr Sebulime retorted. 'Look at Mr Bazanya. He is looking directly into the camera; he is actually posing for the photograph. Look, this photograph is so crowded with people that its educative intention is completely lost. No, we cannot use this photograph to teach other farmers. A good photograph should have few people, at least not more than three, and the pictures should be big enough to be seen', Mr Sebulime confidently asserted.

From the side, Nora, who had only recently joined the participatory development communication research team, looked on in disbelief. The farmer was describing a good photograph as if he had attended a photography class. He was only missing the accompanying jargon of 'centre of visual interest' and 'foreground'.

'They always change what they last said', Moses explained. 'We have spent many months on these photographs and sometimes I wonder whether the production process will ever end. It seems as if we are only going round in circles. Farmers photograph what they wish to appear on their brochures; you take the film to Kampala for developing and printing and by the time you bring the photographs back, the farmers have changed their mind and they would like another type of shot.'

In another corner of the same room, Enock, the agricultural extension worker, looked on as the farmer described the picture he had wanted to capture in the photograph. The group of farmers also listened as Bazanya narrated how a good banana crop should be illustrated:

> *There should be a mother plant with a good banana bunch, a daughter who will take over from the mother, and a granddaughter. The picture should show good mulching practice where you do not put the grass right up to the banana plants. If you put the mulch very near the banana stem, the insects living in the grass will be able to attack the banana plants also.*

Listening to the dialogue, Nora realized that the farmer was doing the extension work instead of Enock, who was only listening. At the end of Bazanya's description, Enock asked him why he had not photographed exactly that. In fact, Bazanya had photographed what he had explained. But when the pictures were brought back for the farmers to choose from, the person who had appeared in that particular photograph had taken it for keeps! So it was no longer available for use in the brochure.

Making posters and brochures

Farmers turned up to make corrections on brochures and posters. It was a good gathering considering the long distances, the difficult terrain and the heavy dust on the roads due to the dry season. Although the farmers had previously been divided into three groups according to their different natural resource management problems, work on the brochures went smoothly, with farmers sharing photos when necessary. For example, if a group's photos did not depict what the farmers wanted to illustrate, the group approached other members for a more suitable photograph. This indicated that the farmers appreciated the fact that the exercise was not intended for competition amongst themselves, but for the purpose of sharing information with other farmers who had not yet taken up the new farming practices. It also suggested that the three initial working groups were slowly merging into one group.

The poster was not as easily constructed as the brochure, partly because farmers said that they did not know what posters look like. This was a challenge considering the fact that the community hall in which we were working had a poster on its wall. There were also several posters in the entrance of the community hall. Upon asking the farmers whether they had seen them, some said no, while others said yes – but that they had not examined them closely enough to know what they were about. When examining one poster that depicted proper water and sanitation practices, some farmers actually said that its aim was to teach people how to write. The concept of poster-making was difficult for the farmers to understand since they were expected to put several photos on one chart, which together would tell a story. They finally agreed to make posters with a limited number of pictures. Initially, the plan was to use still photography, with subjects that the farmers themselves photographed. But certain illustrations proved difficult to capture, such as a hole in the ground that was to be used in the making of organic manure and the trenches that capture soil erosion. In the end, farmers settled for artistic illustrations.

During the material production stage, despite time constraints, the production process appeared to have no end in sight. Facilitators then made it possible for farmers to work with a professional video cameraman and an illustrator in order to finalize the process. This arrangement also solved the problem of farmers taking the photographs for keeps and the limited expertise in the farmer-researcher team regarding the making of posters and brochures.

After the production stage, farmers devised a plan of how they were going to use the communication materials. In a workshop setting, farmers agreed on the geographical scope of their intended farmer-to-farmer information sharing. They confessed that, on their own, they could only manage to share information within their villages. This is because farmers can either walk on foot or ride bicycles to cover the relatively short distances in the villages. They also agreed that in order to be credible to other farmers, they had to make sure that their own banana gardens illustrated the recommended soil management techniques. They identified the different categories of people who needed to be informed about the three modes of soil and water management. These included fellow farmers like themselves, vulnerable groups such as women, as well as the disabled and orphans heading households. In order to get support for their farmer-to-farmer information sharing, farmers included local leaders, politicians and non-governmental organizations (NGOs). They identified institutions and channels through which they could share the information with other farmers. Farmers agreed to make use of radio, a farmers' newsletter and drama to supplement their information sharing through brochures and posters. In addition to one-to-one information sharing, farmers identified village meetings, churches and market days as possible fora. As a way of concretizing their plan of action, they made a time framework within which to accomplish specific activities related to the information-sharing objective. They also agreed on monitoring indicators and mechanisms.

Getting more specific

As we witnessed this participatory material-making process, we realized that there is another possibility regarding material production for the purpose of farmer-to-farmer information sharing. Three sets of communication materials may be produced. One set is intended for the 'teacher farmers' – the farmers who are going to share information with others. This set of communication materials should be illustrative and should enable interaction between the farmers on the identified subject. The teacher farmers should use this illustrative material to explain to the other farmer(s) the recommended soil management practices. This communication material should be produced in fewer numbers and should be durable since it is going to be used over and over again.

After sharing information with the other farmers, the teacher farmer should leave some material with the learner farmer for reference purposes. This communication material should be produced in larger numbers and need not be as durable as the material for the teacher farmers.

There could also be a third set of communication materials that would act as a back-up for the information-sharing activity by providing brief, general information on the subject at hand. For example, this could be a poster that would hang on the walls of farmers' houses, in shops or in churches; it could also be a radio programme.

In the end, it is important to remember that these materials will always be more effective and appropriate if farmers are closely associated with their production or, even better, if they end up taking charge of their production.

Note

1   This chapter was written in consultation with the other members of the research team: Wilberforce Tushemereirwe, Drake N. Mubiru, Dezi Ngambeki, Carol N. Nankinga, Moses Buregyeya, Enoch Lwabulanga and Esther Lwanga.

From Information to Communication in Burkina Faso: The Brave New World of Radio

Souleymane Ouattara and Kadiatou Ouattara

Introduction

In Burkina Faso, as everywhere in Africa, radio exerts an undeniable attraction. In countries where the government has given up its monopoly of the airwaves, nearly all the available frequencies are now occupied by radio stations run by community, commercial or religious broadcasters. Not only do people everywhere listen to the radio, but it has taken root throughout the land, even

in the most remote villages. The success of this tool, which some call 'Africa's internet', allows people to express their concerns in their own languages. Yet, this communication medium remains largely an unexploited resource when it comes to participatory development communication (PDC). An experiment conducted by the Journalists in Africa for Development (JADE) network at three places in Burkina Faso and in Mali demonstrates both the usefulness of such an approach and the importance of taking certain precautions to ensure its success.

From listeners to communicators

Somewhere in Burkina Faso a farmer sits down in a studio and belts out a few songs dedicated to his friends and relatives. The man is about 50 years old. He is happy to be able to make his voice heard in his home village. This is the first time that he has ever entertained a radio audience. Earlier, he had sent his sons with a present of some yams and a chicken for the station technicians.

Elsewhere, members of a rural listeners' club have just delivered their weekly programme sheet to the station. The form is filled out in the national language and shows the broadcasts that they have heard, together with listeners' observations and their requests. The station will take these into account in its coming broadcasts.

Examples of this kind represent amazing progress compared with the way things were once done, from the time the first radio stations were established in Africa until the 1990s, when many countries in West Africa began to deregulate the airwaves. Yet, despite this freedom of speech and the openness that radio maintains with its audience, considerable effort is still required if radio stations are to become true tools for participatory communication. It is all very well to free up the airwaves or to run radio games in the villages; to greet people throughout the day by their own names; to produce and broadcast programmes about farming, livestock and the environment; and to take account of listeners' expectations. All this allows better use to be made of radio; but it is not the same thing as making radio a tool for PDC, and still less for the management of natural resources.

There is no magic recipe for making radio a participatory communication tool. But the experiment that JADE conducted during 2000 to 2002 in three places (two of them in Burkina Faso and one in Mali) provides some food for thought. This chapter looks, in particular, at the most successful effort, launched in January 2001 at Ziniaré in central Burkina Faso.

Central Burkina Faso is a region of poor soils and one that faces severe water shortages. The 2001 project involved using participatory communication to share information among communities, development agents and radio producers in order to improve the management of natural resources. Besides JADE, the group includes technical experts from the agriculture and environment ministries, farmers organized as communication relay points in their communities (also called local communicators) and radio producers. The partner radio station belongs to a farmers' organization, Wend Yam.

Participatory communication: The key to natural resource management

Natural resources are, by their nature, common resources, and their exploitation and (above all) their preservation are in the hands of the community. These resources must be regarded as a fragile asset to be used wisely while being conserved for tomorrow, and protecting them depends in large part upon the community. It is the community who must decide whether or not to preserve forests, water sources and the entire natural environment in order to ensure its own subsistence and that of future generations. Yet, the general tendency is to exploit these resources with a view only to the short term, and there is little thought given to preserving them. We may safely say, then, that there is cut-throat competition between different users (farmers, herders, agro-businesses, etc.) that risks not only mortgaging the future of these common resources, but also creating conflicts over access to them. For all these reasons, natural resource management is a natural target for participatory communication as a means of creating dialogue among all stakeholders and helping them to make decisions that will serve the interests of everyone. Above all, PDC can help to organize joint activities based on a shared vision.

Promoting social groups through radio

It was not by chance that JADE chose to work with radio: it already had long experience in this field. But the first attempts, innovative though they were, did not meet with much success for two reasons. First, JADE was broadcasting with the help of technicians, radio producers and local communicators; but it was not working with the communities themselves, who were treated strictly as sources of information. Second, radio was not used to support participatory communication activities that might help to identify problems, analyse them and find solutions.

Moreover, despite its community radio designation, Radio Yam Vénégré, JADE's major partner in this project, remained a typical rural radio station. It was content to broadcast information to the rural public, the only difference being that it made much use of Mooré and Fulfuldé, the two most widely spoken languages in the region. The results produced by this top-down approach were disappointing and led JADE and its partner to try to use radio in another way.

The new approach recognizes and places value on community viewpoints. The public can express itself directly over the airwaves, instead of simply listening to broadcast advice from experts and radio producers. Discussion groups (for women, young people, adults and seniors), backed up by a team of development experts and radio producers, analyse problems with natural resource management, their causes, their consequences and their potential solutions. This process, which has been recorded from start to finish, involves four stages:

1   identifying problems in natural resource management;
2   programming produced by radio personnel and local communicators;
3   broadcasting the programmes;
4   gathering feedback.

Identifying problems in natural resource management

'What do you consider to be the priority problem with water, wood, fields and the land, and how could you help to resolve it?' This was the first question put to debate in the four discussion groups, and then in a plenary session. The idea was to highlight the community's capacity to find its own means for resolving its problems before turning to other sources for assistance.

At Nagreongo, an arid region of central Burkina Faso, three issues emerged from the discussions: water shortages, soil degradation and lack of wood. A poll was taken, and water shortages received the most votes. This issue was therefore ranked first in terms of priorities.

Among the reasons for the water shortage, people cited the inadequate wells, the great depth of the water table, the lack of erosion control and the absence of reservoirs. A more thorough analysis also revealed that the water was unhealthy for human consumption, and that simple techniques would have to be found to make it drinkable.

In terms of consequences and impacts, it was evident that the lack of clean drinking water led to public health problems, such as Guinea worm disease, childhood diarrhoea and other waterborne illnesses. Moreover, the shortage of water was destroying village communities as their able-bodied workforce gradually left, and was also harming livestock activities.

Programming produced by radio personnel and local communicators

In a participatory approach, producing a magazine-type programme involves not only radio producers, but also local communicators. These are farmers who are responsible for identifying village interests in advance, preparing field trips, taking part in producing a broadcast and gathering feedback from listeners. In this way, a series of 45-minute radio magazine programmes was produced, essentially based on ideas from members of the various discussion groups and interspersed with a few musical interludes. This was 'raw feed' that served to highlight participants' viewpoints.

Broadcasting the programmes

Public announcements were made, usually a week in advance, alerting people of the coming broadcast. This not only made it possible for people to tune in individually, but also, in some cases, to form discussion groups.

Gathering feedback

The local communicators were responsible for sounding out listeners' views on the broadcasts, with the assistance of the listeners' clubs (in Ziniaré and Sikasso). Typically, a local communicator would visit the main regional town on market days, and people would pass on their opinions to him or her. In some cases, listeners with literacy skills would send their viewpoints direct to the station. Generally, the station attempts to respond directly to listeners whenever possible. It also occasionally calls upon resource persons or institutions to deal with listeners' concerns.

Lessons from the experiment

Preliminary steps are needed to establish the sense of trust that is essential in getting the process up and running

Radio stations typically approach the communities with a preconceived idea of a broadcast. People rarely have the chance to make their expectations known in advance. Even today in rural areas, radio stations, despite their proliferation, are still inaccessible to most of the population. The fact is that when people are able to express themselves over the airwaves, this gives them some power, not only within their village but beyond. A good example of this arose in Nagreongo. Radio producers went to the village for a working session with resource personnel in order to establish a basis for collaboration and they found, to their surprise, that the entire village came out to meet them. Yet, this is a common practice, something communities do to stay in the good graces of development agencies. How could the situation be handled? It was here that local knowledge helped to avoid embarrassment. The working session originally scheduled was set aside, and the programme producers set about recording a broadcast on the history of the village, to the great satisfaction of everyone. They then held an interview with the local healer, an influential personality in the village and a great master of ceremony. These broadcasts were carried over Radio Yam Vénégré in the evening.

In this particular case, the radio team turned a dicey situation into a triumph by producing a radio broadcast on the spot. This ability to react quickly to situations allowed them to win the villagers' trust, and then to hold discussions with separate groups of men, women, youth and the elderly.

Collaboration with traditional authorities, technical experts and communities (in other words, teamwork) is an essential factor for success

The participatory approach demands real collaboration among all stake-holders – traditional authorities, research teams and communities – in order to minimize the risk that things will grind to a halt if certain parties are overlooked. Teamwork produces results that reflect the great diversity of the group. In Dori, as in Sikasso, field reporting was considerably enriched by the representatives of farmers' organizations and by rural development experts. The programme producers confined themselves to formulating pertinent questions and ensuring that proper use was made of the recording equipment. Yet, this approach requires a thorough knowledge of the local setting.

Radio gives added value to local folk knowledge

The importance of local folk knowledge has to do with the fact that it is slowly disappearing in the areas under study, without ever having been fully assessed and appreciated. Folk knowledge is not a simple store of information; rather, it has to do with the realm of understanding and of what is sacred. This is why it is important that communities are involved and that the appropriate approach, involving the following steps, is taken:

1 Pinpoint and prioritize the natural resource management problem.
2 Identify the resource persons who have local know-how relating to that problem.
3 Interview those resource persons.

The information collected must deal with the nature of local knowledge, where it was learned, its description, the way in which its holder uses it in practice, the results obtained and the constraints observed.

It is also useful to develop in advance a line of argument to support any information-gathering process in order to overcome the possible reluctance of local knowledge holders and to appreciate their knowledge, which is generally brushed aside by development officers.

That line of argument might point to the fact that as the processors of this folk knowledge move on in years, it becomes ever more urgent to record and preserve what they know and to make it available to the entire population. It may also be useful to point out the paradox inherent in looking to outside knowledge, which is often costly and inappropriate, when there are ready-made home-grown solutions at hand. Local knowledge, indeed, must be treated as a heritage to be preserved.

How can radio help to promote traditional local knowledge? Because it is immensely attractive to the public, radio can give to any information that it broadcasts an importance that it would not otherwise have, not only in terms of building an audience, but above all in terms of winning recognition for it. To be quoted on the radio leads to increased credibility.

The dual nature of radio

Radio is both a means of local communication and a way of spreading information widely. This dual characteristic also makes it a tool for participatory communication and mass communication. Not only does it reach the specific groups with whom the research team is working, but it is broadcast well beyond the bounds of the project's impact area. This concentric spread – the 'oil-spot effect' – gives radio broad scope. A leader of an 11,000-member women's organization in western Burkina Faso tells how the radio helped her in her work:

> *We used to send out flyers to our members to announce our meetings. Some of them were never received. The lucky ones who got the invitations would arrive two or three days after the meeting. Thanks to the radio, we can make an announcement and everybody knows about it at the same time. This has greatly strengthened our organization and its credibility.*

Radio is thus an especially useful medium, both for the mass dissemination of information and for participatory communication. In some cases, it can reach a large audience and amplify messages, while in other cases it allows for real popular participation via the airwaves in the search for solutions.

The question of ethics

Radio: A tool for changing mentalities

Through the new knowledge that it introduces within the community, radio can help to change mentalities. But if it is to do this, there are some preconditions that must be met. We frequently find that broadcasts reflect the viewpoint of the elite rather than that of the majority. The advantage of the participatory approach lies precisely in its capacity to give everyone a chance to express their expectations and their viewpoint, not only to the rest of the community but to people in other villages as well. Radio thus constitutes a powerful tool for changing mentalities, as many examples demonstrate. Women who doubted the effectiveness of vaccinations are now taking their children in for shots. Men who once scorned civil marriage are accepting it today. When it comes to managing natural resources, local knowledge broadcast by Radio Ziniaré is now being tried out in many local villages, where the mistrust that once prevailed has been replaced by a sense of recognition and gratitude.

The limitations of radio

There is a paradox about radio: it is highly seductive; yet it also has the capacity to misinform when it is improperly used. Moreover, it is impossible to interact with the radio host. 'Sometimes it can be hard to understand the message. For example, if someone doesn't get the meaning of a poster you can explain it to him; but with the radio that's not possible', explains one non-governmental organization (NGO) manager in Ouahigouya in northern Burkina.

Generally, subjects of concern to listeners are seldom addressed on the radio. When they are, the treatment may be only superficial. Other technical constraints relate to the station's broadcast range, the relevance of the topics, the cost of producing programmes, their nature, the timing of broadcasts, low audience interest, or the cost of powering listeners' receivers.

When it comes to natural resource management and the participatory approach, we find that radio producers are often poorly equipped professionally, particularly in the techniques of participatory communication. Moreover radio stations, while they may serve a community purpose, tend to become commercial enterprises more attuned to the needs of development agencies than those of the community. The PDC approach, particularly in the area of natural resource management, demands time and skills, which radio stations generally lack. They are more accustomed to entertaining their audience with music or carrying live round-table discussions because they do not have the means to produce programmes in partnership with the communities.

Tools and methods

Discussion groups

Our approach encourages members of the community to participate in debates. Thus, all the selected topics are discussed with the target groups (women, seniors, adults and youth) at the same time. According to Awa Hamadou, a midwife in the village of Kyollo in northern Burkina, 'Dividing people up into groups lets everyone participate in the debates.' A significant step forward has just been taken in this heavily Islamic village, where, in the past, it was only the men who were allowed to speak in public. Now, women are free to express themselves in front of the men on any topic.

In Nagreongo, the research team also relied on four working groups to identify natural resource management problems, analyse causes and consequences, and propose solutions. It is in these groups that the contrasting ways of approaching a problem within the village become clear. It is also in these groups that conflicts of interest are revealed and that solutions respectful of all parties can be found. The advantage of these groups is that everyone has a chance to speak, including women and youth who normally hesitate to express themselves in public.

Audio cassettes

The fleeting nature of oral messages remains a limitation inherent in radio broadcasting. When a programme is carried just once, it has little chance of reaching the entire potential audience. Many programmes are rebroadcast repeatedly to overcome this constraint. Yet, we still find that for many reasons the retention rate for radio information remains low. It is almost impossible to listen to a broadcast several times, as one can do with a cassette. The utility of the radio cassette is that a broadcast can be heard again and again, it can be stored and it can become a tool for sparking discussion about a given topic. Radio stations participating in the project are trading 'magazine' programmes produced and copied on cassettes. Since information on natural resource management issues is not event specific or particularly time sensitive, the audiocassette can be a more appropriate medium than direct broadcasting. Similarly, some broadcasts, particularly those dealing with local knowledge, have been recorded on cassettes to facilitate their circulation to listeners' clubs. Trading programmes among themselves allows individual communities to acquire new knowledge.

Traditional channels: Local communicators

The project on rural communication and sustainable development made some innovations in natural resource management by encouraging reliance on local skills, particularly local communicators. The local communicator was the cornerstone of the project and served as the interface between the relay points (i.e. the project representative in the area) and the community. The local communicator's many functions make him an indispensable player in any action-research effort.

In fact, local communicators perform many tasks: organizing and hosting chat sessions and supporting the research team in identifying natural resource management issues with the community. They also gather local knowledge to be dealt with by the broadcast, identify the resource persons (model farmers), organize recording sessions and gather feedback.

The importance of follow-up

The project's impacts are due, in large part, to follow-up by stakeholders in the field. Development technicians have taken it upon themselves to share experiences, experiments and lessons learned as a way of working with the communities. They can now foster synergy among themselves and, in particular, with the radio producers.

As to the radio producers, they can regularly be found collecting information in the field. They also work with the community in choosing issues, in handling them and in gathering feedback. For this purpose, they have retained the devices that were introduced (e.g. listeners' clubs and local communicators). Today, radio is using the participatory communication approach to go into the field, identify problems and seek solutions with local people. Moreover, thanks to local communicators who have their roots in the community, the issues of concern to local people can be inventoried and, in this way, can form the basis for radio programmes. Local people are now taking a much more active role in producing broadcasts.

In Sikasso, radio producers who were not involved in the project were, nonetheless, impressed by its achievements to the point where they wanted to use the approach in their own work. Within government technical services, as well, several managers recognized the usefulness of radio in dealing with bushfire prevention, sharing knowledge about production techniques, settling disputes over environmental management, preventing disease and reinforcing producers' groups.

Today, Radio Yam Vénégré has more direct and more regular contact with local people and is facilitating negotiation between them and the development agencies. With its experience in PDC, it is now playing an advisory role to development institutions. Moreover, it has helped a project that was designed to strengthen farmers' organizations to clearly explain the objectives of its surveys before beginning fieldwork. Programmes have been produced and broadcast for this purpose, and development officials are now working together. Even outside the project, broadcasts are being produced with other agencies attracted by the approach, something that also earns new revenues for the radio station.

By way of conclusion, we can take satisfaction from having used radio in a new way and we can congratulate ourselves on the resulting achievements. But the research team could have gone even further. It should have systematically produced audio cassettes in order to preserve the chain of communication, to reach those who missed the broadcasts and to allow certain points of debate to be revisited. Posters or booklets would also have helped to reinforce viewpoints. With respect to institutional information – that is, community questions about the project itself and the responses of the project coordinators – use of the radio could have cleared up a number of misunderstandings. This experiment provided a good opportunity for assessing the most efficient and effective types of programming. In addition, PDC could have been compared with the conventional communication format, which is simply based on broadcasting information. Recognizing that the broadcasts occurred both before and after the group discussions, a summary of the previous broadcast could have been replayed before the discussion sessions were held. The radio station, in fact, had the right equipment, including powerful loudspeakers, to perform this task properly.

Despite the importance of rural radio, it should not take the place of chat sessions. Radio should be seen as a supplement, one that relies on the results of these sessions for input. Chat sessions must also meet certain conditions if they are to be truly participatory. In the absence of all these conditions, people will, of course, continue to listen to the radio; but what they hear will be the voices of the producers, the dominant voices of the day – it will not be a tool at the service of the community that is truly participatory.

And Our 'Perk' Was a Crocodile: Radio Ada and Participatory Natural Resource Management in Obane, Ghana

Kofi Larweh

Introduction

Radio is usually seen as the ideal medium to reach people in remote areas who do not have access to other sources of information. But when committed to community participation and people's development aspirations, radio can

be much more than a source of information. It can also be a powerful tool to facilitate consensus-building and decision-making at the community level, thus becoming a catalyst for change. This is precisely what Radio Ada set out to do when it decided to accompany the efforts of the people of Obane in the eastern part of Ghana. After four decades of watching their environment deteriorate, the inhabitants of that community rolled up their sleeves to restore the waterway, which used to be at the heart of their lifestyle, and regain their lost prosperity. Day in, day out, Radio Ada was by their side, voicing their concerns and acting as a morale booster. In the hands of the community, the radio also became a tool for advocacy and mobilization. The results are quite astonishing.

A catalyst for collective action

Many towns and villages seem to have forgotten the reasons that they have come to be where they are. The people of Obane have not.

Obane is a rural community in Big Ada in the Dangme East district of Ghana, about 100km from the capital city of Accra. It is a brisk hour-long walk away on a hot, dusty road from Radio Ada, the community radio station of the Dangme-speaking people.

The people of Obane still remember that the waterway is the reason that their forefathers chose to live there. The waterway, the Luhue River, is a tributary of the mighty Volta River. It supplied fish, provided water for irrigation and served as a bustling transport course. The water was so abundant that Obane used to be the food basket for Big Ada. The women were, among other occupations, fishmongers, farmers, mat weavers and petty traders, while the men fished, farmed or hunted.

That was over 40 years ago.

The creation of the Volta Dam during the early 1960s reduced the flooding of the fields at Obane. Weeds, trees and debris choked the waterway. The surrounding lands became barren, leaving a fetish grove with isolated trees and some patches of green to the south. Not even the shadowy line of emaciated trees tracing the meander of the Luhue River in the background is thick enough to break the line of sight to the west.

Over time, Obane became one of the poorest communities in the district. For this reason, Radio Ada has always taken a special interest in Obane. Radio Ada is Ghana's first community radio station. On the air since February 1998, the station broadcasts 17 hours daily exclusively in Dangme, the language of its listening community.[1] It is staffed by volunteers drawn from the community and trained in its home-grown workshops. Radio Ada's identity is rooted in the culture of its listening community and inspired by their desire to improve their economic way of life while maintaining close community ties. The station actively pursues a participatory development philosophy. One of the main ways in which it tries to operationalize this philosophy is through 'narrowcast' programmes. These regular weekly programmes are recorded in different communities, with the main occupational groups in the listening community – fishmongers, fishermen, women farmers, men farmers and so on. The programmes are driven by the participating community members. They determine the content and it is their voices that predominate. Programmes are presented as a continuing dialogue, where producers simply act as facilitators. The station's holistic approach to community-initiated development as 'the voice of the voiceless' in content and the production process generates trust.

Thus, the programmes serve to affirm the knowledge and experience of group members, who share what they know, what they feel and what they believe. They fuel consultation and interaction between themselves and their counterparts, as well as between other listeners in the Dangme-speaking communities. They also serve to deepen the relationship with Radio Ada. As resources permit, the stations follow up the programmes with more extensive community consultation.

It was after a number of 'narrowcast' programmes had been produced with the people of Obane that one such consultation was initiated by Radio Ada. The consultation was held a few years ago, when the community was confronted with a decision to migrate or to renew their waterway.

The consultations generally involve the use of participatory rural appraisal (PRA) tools, such as the ranking-and-scoring matrix for needs or priorities. Despite the inherent friendliness of the tools, often when one asks communities about their priorities, the answer goes something like this: 'We need pipe-borne water, roads to transport our produce, electricity, employment, etc.' The litany never seems to end and tends to read like a shopping list.

In the case of Obane, possibly because of the ease and trust that had been built in their relationship with Radio Ada and almost certainly because they were at a decisive point in the life of their community, the responses were more detailed, reflective and textured. In every case, respondents had a story to tell connecting the present to the past.

The fishermen lamented that they had become goat and sheep rearers; one owns a cattle kraal. They were worried that the different fishing skills they had learned from their parents and friends were no longer of value to them and their children. The river did not exist during most of the year. It was choked, and even during the short rainy season it provided only limited access. The vast marshy lands that had boasted of crabs and reeds for weaving had all run dry.

The women could not bear sending their children off to school in Big Ada and beyond without enough food to support their stay. Formerly a little rise in the tide caused natural flooding of the fields for fresh crops even during the dry season. Although Obane lacks many basic amenities, the women identified 'unity' as their most important priority for development. They were sure that they, working with their husbands and joining other neighbouring communities, could revive the economic life of the community and restore it to its former status as a food basket.

During the plenary discussion, Divisional Chief Nene Okumo was invited to share his reflections. He narrated how during the past their fathers occasionally cleared the waterway to support ecological processes and to sustain flora and fauna. Then and there, the plenary decided to dredge the Luhue River using communal labour. Immediately, Madame Adjoyo Djangma, a mother of seven and redundant fishmonger, intoned a jubilant, melodious traditional chorus. The response was infectious as others took up the chant.

There were contour lines on some foreheads – more than 40 years of vegetable growth and silt to clear! The work extended over 10km, and would entail varying degrees of difficulty. In the meanwhile, the other neighbouring communities had their own priorities and there was the need for tools and other logistics. How could they possibly mobilize the resources needed?

'How can your community radio station, Radio Ada, help?' There was a measure of relief on some faces as the role of communication was put forward: 'Announce what we have decided to do ... announce the day and time for the work ... when others hear, they will join us ... announce the names of those who have reported for work ... put our needs on the radio ... tell other people about the by-laws we shall make to prevent the river from choking.' The ideas kept coming.

Having facilitated consensus- and decision-making on the ground, Radio Ada fuelled the communication bonds within the Obane community and connected them to other communities and institutions to support the communal labour needs. Trusting the station to honour its offer of support, the groups actively turned it into a tool for advocacy and mobilization. They announced at their convenience and free of charge their work plans – whose turn it was to work – and raised issues on air about the environment, their occupations, their lives and the project's progress.

Both the women and the men of Obane conducted the clearing, and 60,000 seedlings of red and white mangrove have since been planted.

Before long, four groups from different communities – Obane, Gorm, Togbloku and Tekperkope – were working together to dredge the Luhue River. Other groups from other communities joined them to show solidarity and were duly acknowledged on air. They included communities from Dogo, Dorngwam and Atortorkope, as well as from Aminapa, Wasakuse and Midie.

District Wildlife Officer Dickson Yaw Agyeman stated: 'The radio broadcasts were a morale booster and a challenge. The Wildlife Department gave them Wellington boots, nylon ropes, cutlasses, hoses for weeding and food for work – kenkey[2] and fish and pepper.'

Their voices and active participation on air also won them other collaborators, allies and supporters. Organizations and institutes such as the Dangme East District Assembly, the Canadian High Commission, Green Earth and the Kudzragbe Clan of Ada Elders contributed to the initiative.

The men's group received a loan facility of 14 million Ghanaian cedis[3] to boost agriculture, while the women obtained a loan for 15 million cedis for food processing and marketing. In addition, a 12-seater KVIP toilet[4] was built. Funding was through the Wetlands Management and Ramsar Sites Project, which is supported by the World Bank. The magnitude of work accomplished over these four years might intimidate a stranger; but knowing what the river and the surrounding fields mean to their lives, the people of Obane persisted. With the support of other communities and of the Wildlife Department, as well as their community radio station, Radio Ada, they now have what matters most to them.

Day in, day out, as needed, Radio Ada followed the efforts of Obane on the air. Throughout the process, it took its cues from the leadership of the community, who were constantly at the station making requests for specific programmes and announcements. Radio Ada not only helped to mobilize concrete material support for the back-breaking work of the people of Obane, but also inspired the rest of the community with their enthusiasm, perseverance and sense of unified purpose. By consistently projecting a community endeavour into the public domain, as well as through the soft cheerleading style of the radio broadcasts, even disputes over land boundaries, customs and leadership were avoided. No one could afford to risk name or reputation by swimming against what was now literally a growing tide.

Today, one can travel on a boat from Big Ada through Luhuese to Obane and beyond. Year-round small-scale farming is possible by using irrigation methods, and the people now have access to freshwater. A replication is also in the offing at Totimekope near Ada Foah, the administrative centre that is the twin town of Big Ada. There, the Futue River is to be cleared in order to restore river transport, fishing and farming. The satisfaction of Radio Ada is in fulfilling its mission and sharing in the joy of the community's successes. Occasionally, however, there are more concrete rewards – perks, so to speak.

One sunny day, one of the dredging groups, led by Alfred Osifo-Doe (a professional mason), burst straight from the dredging site, sweating and jubilant, into Radio Ada, excitedly bearing a gift for the station – proof of the success of the common endeavour, they cried, fished out of the increasingly swelling waters of the Luhue.

Deeply touched, but mindful of its conservation role (as well as of the practical difficulties involved), the volunteers of Radio Ada managed to persuade the dredging team to return the gift to its natural habitat. So, be careful the next time you cross the Luhue River because by now it must have grown – the baby crocodile, that is!

Notes

1   The station has about 600,000 listeners, within a 100km radius. Although there are variations in the Dangme language, such as Klo, Gbugbla, Se, Ningo and Ada, the people are bound by a common history and culture. The station derives its name from the location in relation to the language. It is based at Tetsonya near the main town, Big Ada, a two-hour drive from the capital of Accra. The main towns in the radio's coverage area are Ada Foah, Big Ada, Kasseh, Sege, Goi, Akplabanya, Otekporlu, Agogo and Asesewa. The Atlantic Ocean and the Volta River form part of the boundaries of Dangmeland.

2   One of the local staples and delicacies, 'kenkey' is made from fermented maize dough rolled into balls and boiled in dried maize husks. It is usually eaten with

charcoal-grilled fresh fish and a relish of freshly ground hot pepper, onions and tomatoes.

3   At the time, US$1 was worth approximately 6500 Ghanaian cedis.

4   Kumasi ventilated improved pit (KVIP) toilets are an improved version of pit latrines.

Burkina Faso: When Farm Wives Take to the Stage

Diaboado Jacques Thiamobiga

Introduction

In some parts of Burkina Faso, women do not have the right to speak in public. Yet, they play a key role in the development of their communities. As farmers, they can observe the changes that are taking place in their environment, such as the impoverishment of the soil in the fields where they work. But because they are excluded from public debate, it is hard for them to help find solutions to these problems. In the case study related in this chapter, the women of two

villages in western Burkina Faso turned to their local traditions and culture to find a way of speaking out and launching discussion on problems such as soil erosion, soil productivity and even the property owning rights of women. This is a fine story to tell a cousin who will be glad to know that in his native region, participatory theatre has allowed women to address these questions publicly and, at the same time, to raise their status within their communities.

Letter to a cousin

Dear Cousin:
In this letter I want to tell you the story of the Burkina Faso farm wives who produced a first-class theatrical piece. As you know, people who have been to school think that those who haven't are unable to mount theatrical productions of the same quality that they can. Farm wives in the villages of Badara and Toukoro, in Burkina Faso, have now given the lie to this assumption. They never went to school, and they can't even read and write. Yet, they succeeded in putting on a truly high-quality 'debate theatre' production.

They were helped in this effort by a multidisciplinary team consisting of a sociologist, an agronomist, a theatre producer, a communicator and a video producer. With this support they were able to identify the problem that was to be the subject matter of the play, which they then created and presented in the villages of Badara, Toukoro, Tondogosso and Dou. After the show, the audience offered some criticisms, as well as some ideas for improving the play. Finally, the women drew some lessons from their experience.

The problem targeted by the play

The story I am going to tell you took place in two villages of western Burkina Faso, Badara and Toukoro. For your friends who are unfamiliar with Burkina Faso, you can tell them that the country is located in the heart of West Africa. It is one of the poorest countries in the world; indeed, the United Nations Development Programme ranks it 173 out of 175 countries in its development score.

Burkina has no oil or diamonds. Its only wealth is to be found in its men and women, who are dedicated to work, especially on the land. Its people depend upon farming and livestock for their livelihood. In the past, the harvests were abundant. Burkina villages hardly ever went hungry. This is not true anymore. The land no longer yields good harvests for it is worn out and impoverished. As if this were not bad enough, the rains are unevenly distributed over space and time. It's as if Mother Nature were angry with Burkina and its people.

That was the case this year. Many farmers have had no harvest from their fields. One of them said the other day on television that he had not harvested 'a single grain of millet'. The women in the two villages that participated in this project said the same thing on many occasions. The president of one of the women's groups in Badara told us: 'The village's poor harvests are due to a lack of soil fertility. The situation is so severe that the village is going to disappear.' She added that 'You don't need a degree from an agricultural school to know that the soil in the fields is no longer healthy. All you have to do is look at the stalks of millet or sorghum.' One of her friends said the same thing, adding that 'the land is worn out' and 'the fields are full of striga', a weed with violet flowers that springs up in fields that have lost their fertility. The words of these two women were backed up by other women in Badara and Toukoro. The team agronomist confirms that what the women say is true. The soils around these two villages are not as fertile as they were, although they have not deteriorated as far as those in other parts of the country.

The women point out that they do not often have the opportunity to talk about this problem with each other, and still less with their husbands. Most of the time, extension workers provide technical advice only to the husbands, who may not even talk about it with their wives. The women have no radio stations where they might get advice. Furthermore, they are not allowed to talk about the problem in public, for their husbands will say that's none of their business and may even get angry with them.

What we have here, then, is a big communication problem. We can say that through this theatrical experiment the women have sought to discover how they can address the problem of soil fertility so that their villages will understand that something has to be done about it.

The creative theatre process

When the women decided to take action, they asked themselves how they could put the problem to their villages without making their husbands angry. It was through discussion amongst themselves and with the team's theatre producer that the women of Badara were reminded of a traditional ceremony where they are allowed to dress up as men and speak a few plain truths, and the men have no right to be annoyed. It is village custom that gives the women this privilege, and it is binding on everyone.

The ceremony is held whenever the rains dry up in the middle of the rainy season, thereby threatening the annual harvest. During the ceremony the women appeal to the bounty of nature. They insist that when they come out in this disguise to make their supplications to heaven, the skies will open up and will send them home in the rain. That was the idea behind this debate theatre production. Members of the theatre troupe helped them to put together the production in six steps, which I am going to describe for you.

Step 1: Getting to know the villages

The women set out to see whether the debate theatre idea would be acceptable in their villages. As you know, nothing can be undertaken in the village without the approval of the customary, religious and administrative officials. That's why you have to know about local customs, what is allowed and what is not.

Moreover, mounting the production requires the consent of the husbands, without whom the women would have been unable to participate.

In the end, the women turned to their knowledge and their know-how concerning soil fertility, as well as communication. This step, which was assisted by the team sociologist, allowed everyone to understand the villages more thoroughly. It also helped to establish relations with everyone. The women were even allowed to move to another village with members of the theatre troupe for three weeks. Every Saturday evening they would go home to see their families, and on Monday morning they went back to their temporary quarters.

Step 2: Preparing the agronomic model

An agronomic model was then prepared with the help of the team agronomist. The women inventoried and analysed their knowledge and their know-how concerning soil fertility and they defined four key principles:

1   The land feeds the people, so the people must feed the land.
2   If you want to feed the land properly, you must understand it thoroughly.
3   If you use only inorganic fertilizer on the land, it's like putting water in a basket.
4   If you want the land to feed your children and your grandchildren, start taking care of it now.

The agronomic model includes techniques that allow producers to:

*   combat soil erosion by building anti-erosion installations such as stone retaining walls and earthwork dikes;
*   protect the soil against wind erosion and sunscald by mulching;
*   prepare the seeding bed using the traditional zai technique, which involves digging small pits and putting organic manure in them, and then placing the seeds in them when the rains come;
*   preserving the fields' existing tree cover, at the rate of 25 trees per hectare (the general standard);
*   enriching the soil with organic manure at a rate of 2.5 tonnes per hectare (the general standard);
*   combining organic with inorganic fertilizer; and
*   combining different crops, such as cereals and legumes.

Step 3: Conceiving the play

Using the agronomic model and collected materials on patterns of communication in the villages, the theatre troupe representative created a play called Sétou's Challenge. This play tells the story of a village woman who was deeply concerned about the loss of soil fertility and about the tough living conditions of farmers, men and women alike. She took advantage of a big celebration in her village (a child's baptism) to talk with other women about the problems affecting them and their village.

It was during this celebration that a strange character named Doda, the village fool, appeared. To everyone's surprise, he began to dance with the women; but they shoved him away in fear. Seizing the occasion, Doda suggested that they do a play. At first, the women thought this was a silly idea; but then they listened to him more closely and accepted his challenge. They decided together to employ the device of debate theatre to address the problem of soil fertility that was impoverishing the women and their village. As Doda put it: 'Debate theatre is a game that not only amuses people, but makes them think about the problem of soil fertility.'

Step 4: Creating the play

This was the most interesting and, at the same time, the funniest step. It was, in a sense, a theatrical piece in itself. Creating the play involved multi-sided negotiation among:

- the women themselves within their associations, and between the women of the two villages (Badara and Toukoro), for selecting the actresses;
- the technicians, the husbands and the village authorities, on the one hand, and between the women and the technical members of the multidisciplinary team, on the other hand;
- the technicians of the multidisciplinary team.

In this way, 14 women (seven for each village) were selected to act in the play. In order to bring the women from the two villages together and to give them a couple of weeks to put together the production, it was decided, with the agreement of the husbands, to move the troupe to the rural activities centre of the village of Banakélédaga, midway between the two villages, where a theatre school was set up. It took some patience for the women to learn to be comic actresses, and they worked at it for four weeks with the help of the playwright and the two stage directors. Everyone approached the task with true professionalism; as a result, the women were ready to put on their play after four weeks.

Step 5: Presenting the play

When the play was ready, the women gave five performances (two in Badara, one in Toukoro, one in Tondogosso and one in Dou). Every performance was a big hit. It was especially fun for the women when they got to do the play in their own village; whenever they came on stage, the local audience was astonished to see their mothers, sisters or wives dressed up as men. Many husbands recognized them by their gestures and their manner of speaking. In fact, each performance sparked a big celebration in the villages.

Step 6: Evaluation

The debate theatre experiment was evaluated using several procedures. During the performances, and especially during the discussion that followed them, members of the audience offered some criticisms and proposals for improvement, and these were subsequently acted upon. In addition, the actresses held discussions after each step among themselves and with the multidisciplinary team about the accomplishments, the limitations and the difficulties of each phase. The multidisciplinary team, for its part, assessed each step in the process as it unfolded. Finally, a theatre expert conducted an external evaluation of the whole exercise.

Results of the theatre experiment

Through this debate theatre experiment, the women created a good atmosphere, that of a public celebration, within the villages where they gave their performances. Seeing the women dressed up as men made people laugh, of course; but it also made them think about the problem of soil fertility. One member of the audience told us: 'Seeing the women in costume set me to thinking more than laughing.' One woman expressed her thoughts to us in the following terms:

> *By disguising themselves, the women were portraying a real-life situation; they are often called upon to fill the role of absent men. This is no longer a disguise but a reality that we have lived. Under these conditions, the women are no longer just wearing men's clothes; they are also taking charge of their families, something that used to fall to men, just as widows must do.*

Next, the debate theatre served as a learning experience for everyone: men, women and children. People exchanged viewpoints about village life, they discussed male–female relations, and they shared technical information for improving and preserving soil fertility. Can you imagine? Farm wives giving technical advice on soil fertility to men! This was a first for our region.

Finally, there was a great turnout by people in all five villages where the women put on their play. In every case, there were at least 200 people – men, women, children and young people – in the audience.

There were, indeed, many other results, of which we may cite only a few:

- People responded enthusiastically to the play, making it a real tool of participatory communication for development.
- Villagers became more aware of the need to preserve and improve soil fertility.
- The social status of women was raised within their villages.
- Uneducated and illiterate women developed the capacity to speak in public through the device of the debate theatre.
- Members of the multidisciplinary team strengthened their skills.
- Tools were developed for reproducing this experiment in action research elsewhere and for capitalizing on its output (video, documentation and technical reports).

In the end, the play had some very positive impacts on families and on the villages. On the agricultural front, the people who took part in the debate theatre have put into practice some of the new techniques that they saw demonstrated there. For example, manure has become hard to find in the villages where the play was produced because farmers are now using it more frequently on their fields.

As well, the zai technique is now being used in their common fields by the women who took part in the play. People often ask the women to do the play again. A survey in the two villages (Badara and Toukoro) showed that the play had helped people to find solutions to soil fertility problems, and to overcome inequalities between men and women.

When we say that Burkina's lands are poor, this is both true and false. It is true in the north of the country, where most of the soils have become lateritic. But it is not true in the west, where the soils are still relatively fertile. Badara and Toukoro are located in this part of the country.

For that reason, the knowledge and know-how of southern women was limited to the use of chemical fertilizers. On the other hand, women from the north had experienced the acute problem of collapsing soil fertility in their native area. They were able to set an example, then, for using household wastes and tree branches to contain rain runoff, planting crops of niébé (cowpeas) to fix soil nitrogen, and using techniques such as mulching, zai and stone retaining walls. All these techniques were confirmed by the agronomist and by the theatre expert in the course of supplementary research in Kouni, a northern village where the soils have become highly degraded and infertile.

Yet, it is in terms of gender relations that the debate theatre experiment produced the most surprising results: by letting women speak out, it revealed some unsuspected aspects of the soil fertility problem. In fact, far from ignoring the importance of regular fertilization of the soil, through their stage characters the women explained that without property rights they had no interest in investing to improve yields from a piece of land that the men could simply take back once it had become productive again.

Lessons drawn from the theatre experiment

One of the lessons we can draw from this experiment has to do with the basic problem: we must undertake an in-depth study of villages in order to understand them properly. Research and development projects tend to downplay this phase – the local setting assessment – despite its importance. Financial partners often dispense with it as too costly. Yet, such a study improves our knowledge of the social and cultural habits of villages, their agronomic aspects and their communication practices.

The experiment also revealed the importance of involving local people from the outset in defining and analysing the problem because outside experts do not always see the problem in the same way as the inhabitants who have to live with it on a daily basis. On this point, the example of the women who worked the property ownership issue into the play is highly revealing.

In addition, the turnout in the villages where the play was performed shows that debate theatre can be an effective tool of participatory development communication. It can also facilitate dialogue between women and their communities on a given development problem because it is well suited to the realities of rural life in Burkina. It can thus allow women to play a greater role in developing their communities. This is a tool, then, that can rally different social categories (women, men and youth) and different occupational classes (farmers, herders, etc.) to the cause of developing their communities.

The experiment also showed that illiterate farm wives can accomplish great things if we will just believe in them. This is why it was essential to have them involved throughout the process. The play offered the chance to share the knowledge they had, and to acquire new knowledge, while releasing them from certain social and cultural constraints so that they could speak in public. They even ended up giving some advice to the men! This does not happen often in village communities, which are still in thrall to custom and tradition. If you should run across some of the women who took part in the play, you won't believe that they are simple farm wives. Some of them have had the chance to participate in workshops where they express themselves just as surely as the experts. In other words, debate theatre is an instrument for raising the status of farm wives and winning recognition for them as individuals, as much as for their knowledge and their know-how.

On the institutional front, debate theatre can create a setting conducive to effective partnership if each partner agrees to play his role openly and honestly. It provides the opportunity to strengthen institutional capacities through the effect of complementarity and synergy of efforts. It is essential, however, that institutions involved in the process have a proper understanding of their respective attributes from the outset.

Some criticisms of debate theatre

There is a village saying that the onlooker can dance better than the performer. In other words, when you are in the position of an observer, you are well placed to offer sound criticism. Similarly, when you have done something yourself and then look back on it some time later, you can also make sound criticism. By standing back a bit, we can offer some criticisms about this debate theatre experiment.

First, setting it up is a long and difficult process. That's mainly because of the approach taken for conceiving the play. Moreover, the play was first written in French by the theatre troupe manager, using material he had collected from the women. The play then had to be translated into Dioula, the national language. Yet, it could have been written directly in Dioula, using the women's own expressions, which could then have been more readily reflected in the text. Moreover, since the women were, for the most part, illiterate and had never performed a play, it was hard for them to learn their parts by heart. It would have been useful to work out ways of writing plays together with the women, and this would at the same time have helped to give greater recognition to their knowledge and their know-how.

Nor was the process facilitated by the decision to create a single theatre troupe involving women from the two villages, and to house them in another village half way between the two. This meant negotiating the husbands' consent and knocking off work every Saturday so that the women could go home to see their families. In the village, as you know, the woman is everything to her family. It's hard for her to leave her family to go and work in another village. That's why one of them quit the troupe at the beginning of the production, and she had to be replaced.

Next, the process was expensive. It took a lot of money, especially when several experts were brought in. It would have been a lot cheaper to use a village troupe that is used to putting on performances without great financial backing.

I must also point out that the team consisted of members from three different institutions,[1] and this was another source of difficulties, especially since they did not all live in the same city. Because they were located so far apart, it was not possible to stick to the work schedule, and things tended to drag out. While the play was supposed to be produced after six months, it actually took two years. It would have been simpler and more efficient to work with a multidisciplinary team from the same institution, with the time savings that would have implied.

Finally, the women and grassroots communities showed that they were ready and willing to commit themselves. But they were somewhat disappointed by the fact that not all the needs raised by the play could be met because of the lack of physical and financial resources. Moreover, the team agronomist has not been able to follow up with the farm wives who have been experimenting with the new techniques for soil fertility preservation that were addressed in the play.

Conclusions

In reading this story, you must have realized that this debate theatre experiment served to win new recognition and appreciation for the knowledge and know-how of Burkina Faso farm wives relating to soil fertility. Even better, they enhanced their own self-esteem by showing the men that they were quite able to discuss their communities' basic development problems in public through dialogue and participatory development communication. Debate theatre is, thus, a tool that people can use for dialogue about the development problems of their villages. Finally, it can be an instrument for rallying people around the villages' development efforts.

That's why I want you to recount this story to your friends. Above all, tell them that debate theatre can be the yeast that leavens the dough of participatory development communication. By doing so, you will be helping to put to use the wonderful work of the Burkina farm wives who experimented with that debate theatre production.

That's all for now. Goodbye and take care!

Your cousin from the village.

Note

1   The Centre d'Études Économiques et Sociales de l'Afrique de l'Ouest (CESAO), based in Bobo-Dioulasso, the Théâtre de la Fraternité, based in Ouagadougou, and Zama Publicité, with offices in both cities.

How the Parley Is Saving Villages in Burkina Faso

Diaboado Jacques Thiamobiga

Introduction

Because so many African villages have retained their oral traditions, ancestral forms of communication based on dialogue can be very useful tools in participatory development communication. Yet, the younger generation does not always keep up these traditions and practices, even though they are often better adapted to the local setting. In the process of rural development, they encourage people to speak their minds, they facilitate consensus and they promote joint endeavours. Here is a letter to a cousin, dealing with one of these forms of communication: the parley, or, as it is known in Burkina Faso, the palabre.

Letter to a cousin

Dear Cousin:

Do you remember the phrase that our grandfather liked so much and that he kept repeating? He would say that 'the parley is what saves the village'. When we were little, we could not grasp the wisdom of that saying. Today, I'm going to try to explain it to you through two stories. One of them comes from a journalist friend who describes how the parley allowed the village of Silmiougou to put an end to its water wars. The other story has to do with ten villages in eastern Burkina Faso that used the parley to control bushfires. To help you understand the stories, let me tell you first about the parley and the role that it has always played in our villages.

Talking eye to eye

When we were children, you'll remember, we often saw the elders seated under the big tree in the village. We would ask our grandfather why the older men got together and spent their time talking as if they had nothing else to do. He would say: 'It's the parley that saves the village.' He would always tell us the same thing, that the elders were looking for ways to solve the village's problems through the parley, the African-style town hall meeting. He would finish by telling us: 'Parleying, it means looking each other in the eye or speaking a few plain truths.' In fact, grandpa was right. No problem can be solved unless the people of the village are prepared to sit down and discuss it, talk it over. That's what the parley is all about.

You see, parleying and discussing mean the same thing. But a village parley doesn't happen just by chance. Remember how grandpa and the other elders would bring no one but their grandsons to the meeting and let the kids play while the elders parleyed? It was only the heads of the families and clans, only the men with beards and white hair, who were allowed to participate in the parley over village affairs. You understand: they couldn't just let everyone into the parley when they were talking about witchcraft, death, the rape of young girls and sensitive things like that.

The parley would not end until the village elders had arrived at solutions that everyone could accept. There had to be consensus. And those solutions also had to fit with custom. It's not always easy, as you know, to find solutions that can command consensus and are consistent with custom. That's why the parley often ran on for a very long time. Everyone had the chance to hear what the others had to say, and then to have his own say. So the parley was an opportunity for dialogue. Everyone could express himself freely; but everyone was always respectful of the others and of the village's customs.

Today, we can say that the parley is a traditional way of establishing communication, and one that is particularly well suited to our villages. This approach fits very nicely with participatory development communication (PDC), which is based on both participatory processes and on traditional or modern media, as well as on interpersonal communication and facilitation skills. The idea is to find a solution to development problems as identified and defined by the communities themselves. For those of us who did not go to school very long, this means that the whole village has to get together and discuss things in order to find solutions to problems that burden our village life.

With this approach, the village can:

- pinpoint the development problem that has sparked us to use participatory communication;
- identify the individuals or groups affected by the problem;
- define the needs, objectives and activities for participatory communication;
- select the channels, means and tools for participatory communication;
- test out those channels, means and tools;
- use those tools in the development process, once they have been tested;
- assess the outcomes at each significant stage.

Today, many development partners supporting our villages are using PDC. I'm going to tell you about two such cases, one of which had to do with managing conflicting water uses, and the other with the better handling of bushfires. These two initiatives both relied on the parley, although they modified it a bit. For example, the parley is now open to all social categories (men, women and young people). And it also uses new communication tools such as film, video, radio and cassettes. Let's see, then, how the parley put an end to the battle over water in Silmiougou.

Ending the water wars in Silmiougou

The elders of Silmiougou will tell you that the village was founded by the Peuls. They lived by raising livestock, which at that time was a flourishing activity. But it also attracted many rustlers, thieves who would come into the village to steal sheep, goats and even cattle. Fed up with this thieving, the original Peul settlers turned for help to the grand chief of the Mossi, Moogo Naba, who lived in Ouagadougou. He sent them a band of warriors who were used to dealing with rustlers, the Tapsobas, and they succeeded in driving the bandits out of the village. The Peuls could now live in peace.

However, the Mossi who came to protect the village decided to stay on to keep the village secure. Little by little their numbers grew, and they took over the local lands, including those on which the Peuls pastured their animals. This plunged them into permanent conflict with the Peuls, a conflict that the people of Silmiougou came to call 'the water wars'.

Not long ago, whenever the women went to draw water from the wells, they would get into a spat. The Peuls also had to fight to have their animals drink at the wells. Silmiougou was a daily battleground. To solve this problem, which could have ended in a real war within the village, people decided to turn to the team which was working with villages to stop water disputes. The project fielded Bila, who, instead of acting as a coach or extension worker, decided to use the traditional parley to bring the people of the village together to discuss ways of using the well without squabbling. In this way the village people were able to:

- think about how everyone could use the well water without fighting;
- exchange ideas about all the problems that could lead to water disputes;
- create groups to find solutions that would suit everyone;
- agree on solutions and how to apply them to everyone, even the village chief;
- apply those agreed solutions to everyone and enforce them on all water users; and
- hold frequent meetings to see what was working and what was not, and to find new solutions.

This is how the parley ended the water wars in Silmiougou. As you see, grandpa was right when he said: 'It's the parley that saves the village.' Thanks to the parley, the village of Silmiougou avoided a war over water. This story, told to me by my journalist friend, reminds me of another one about ten villages in the eastern part of the country where the parley made it possible to put out bushfires.

The bushfires are out

As you know, Burkina life depends upon the bush – that's where we find river water, wild animals, new farmland and trees. Women cut wood in the bush for cooking. They also collect the leaves and fruits of certain trees. And it is in the bush where livestock graze on grass, leaves and twigs. So anything that affects the bush in Burkina also affects development.

We must admit that people sometimes act as if they are unaware that the bush can be damaged and degraded. This is what happens with wildfires that destroy everything in their path. As you know, bushfires are frequent in Burkina. For various reasons, people make great use of fire during the dry season. Some claim that this is a traditional farming practice. But the problem is that some fires get out of control and lay waste to the whole area. This has become a real scourge that our country must address. That's why, in 1997, Burkina held a big meeting – a national forum – on fires. That meeting led to an initiative to promote sound fire management. Between 1999 and 2003, the initiative reached about 255 villages. It did so primarily through the parley.

In this case, the parley involved everyone in the village (men and women; young people and old; farmers, herders and merchants). It set up a permanent dialogue among these groups, and also with the technical support services in environment and agriculture. This is what happened with the ten villages that I'm going to tell you about. We discovered those villages during our research into the problem of bushfires – their causes, impacts and solutions – and the way in which people in these villages were handling fire management.

In this story, we will be talking about fire management committees, their work and their results. At the outset of the project, every village organized a fire management committee through a parley in which everyone participated. The committee has ten members: eight men (adults and youths) and two women. The committee's main role is to mobilize the entire village around fire management activities.

These activities involve selecting a fire management site, using controlled 'early burn' techniques,[1] protecting the site against wildfires and planting trees. All these activities are planned, implemented and assessed by the village population as a whole. Every social category has a role to play. The young people are responsible for surveillance and early burning. The women bring food and water to the young people working on the site, and the older folks provide advice to the youngsters and women.

As you can see, these activities could never happen without the parley, which takes place in the form of a general meeting of the village. During our research, we attended several general meetings. Let me tell you about one in the village of Kiparga, which we selected by chance. On that day, 70 people – men, women and young people – were in attendance. The participants expressed their viewpoints and exchanged knowledge about fire management, about the running of activities and about the problems encountered. Sometimes the parley can become very heated, and it is at this point that the elders intervene with words of advice for everyone.

Occasionally, to lighten things up, some cousins à plaisanterie[2] will jump in and get everyone laughing. They make jokes at the expense of those whom they are allowed to tease. They insult each other as if they were going to come to blows. Those who are unaccustomed to this practice don't understand it. They wonder how people can attack each other in this way and yet no one is allowed to get angry. This is what keeps the meetings interesting and fun, even if it means they may run on for a long time.

Things also get very lively at the meeting when the women use jokes to tell the men a few plain truths. Sometimes the very purpose of the meeting can be obscured by laughter. This gives an idea of the spirit and the philosophy that underlies the parley. It is a freewheeling and democratic debate. We could even call it a people's assembly. Thanks to the parley, people can plan their fire management activities: surveillance, protection of the bush and exploitation of its products. Working in this way, the villages have achieved real results and people are proud to point to them. People have rallied around the village efforts at fire management and they are committed to a process of permanent dialogue on this issue. The interests of the various social categories and occupational groups that make up the village can now be taken into account. Not only have the villagers managed to agree on the appropriate ways to manage bushfires, they have also launched other activities to protect natural resources, such as setting up and maintaining protected areas[3] or improving their herding methods.

Yet, there have also been some problems. Heavy pressure on natural resources, due to the demographic explosion, is one of these problems, as is the unconstrained exploitation of natural resources. Moreover, the villagers may be too busy with other activities to keep a permanent watch over the land. These problems, taken together, are causing the gradual disappearance of many useful plant species. Finally, because the village does not have a flour mill, it is sometimes hard for the women to play an active role on the committee.

Despite these difficulties, the people are emphatic about their determination to pursue bushfire control through surveillance, building gravelled firebreaks, site maintenance and planting trees at the site and in the village.

The parley: Advantages and demands

Each member of the village can see advantages in the parley. In fact, the parley affects a great many people through the general meetings and other assemblies. It is also a place for sharing ideas and experience, knowledge and know-how. It offers people a chance to think together about their common problems and to talk them over face to face. It is a means of dialogue and of negotiation that allows people to reach agreement on what they are going to do together. And, finally, if it is properly organized, it does not need much in the way of materials or money. All that's required is a meeting place and some mats or benches to sit on.

We must recognize, however, that there are some conditions that must be fulfilled. In the first place, someone has to take care of the town crier who calls people to the parley by buying him his cola[4] or his dolo.[5] Second, the facilitator has to be someone who takes the job seriously and enjoys everybody's respect, or he will never be able to guide the debate. And then, too, the elders have to be in attendance because they have a calming effect on people when the debate gets too heated.

On top of all this, there has to be a little money to buy food and refreshments for people if the parley is going to be a long one. Holding a good parley does take time, after all. Above all, people have to be patient. If you're planning to start the parley at 9.00 am, it will likely get rolling around 10.00 am or maybe even 11.00 am since everyone will have something to do at home before coming to the meeting. Then, when people arrive, it will take at least 30 minutes to greet each other properly. And each speaker will have to start with the customary string of salutations. Indeed, the parley takes a lot of time, and this can be a problem. The fact is that some people are now insisting on being paid before they will come to a parley, complaining that they have to leave their work to do so, and that 'parleying doesn't fill anyone's belly'.

It is also true that people may be called to many parleys. The extension worker will have his parley. The forester, the prefect, the deputy and the village chief will have theirs too. People may get fed up with all these meetings – after all, that's not the only thing they have to do with their time. For this reason, a parley will sometimes attract only a few people. These days, if you invite 100 people, you can be grateful if 30 actually show up at the meeting. Finally, people sometimes come to the parley thinking that the conveners are going to offer some tangible assistance to the village, whereas it is the process of dialogue itself that will eventually help the villagers find solutions to their own problems.

We must also recognize that the parley is one of those rare occasions when people can express themselves openly. Half of the available time may well be spent on issues other then those for which the parley was called. As you know, people do not all think the same thing. Often, each person will try to show the village that nothing can be done without him. The parley can become impossible if people from different districts are divided by long-standing rivalries and can agree on nothing. In one of the villages we visited, for example, there are two chiefs. You have to be a chameleon, or at least a true diplomat, to adapt to village realities. Otherwise you will just stir up conflicts and you will have everyone against you.

Under these conditions, it isn't easy to guide debate within a big group. You have to know the people and their natures. For example, when an elder speaks, he's going to start by retelling the village's history, before he comes to the point he wants to make. He may even speak in proverbs – and when someone speaks to you in proverbs, you have to respond in proverbs. Then there are the long customary salutations that I mentioned earlier. So, you see that the parley is a time-consuming process. And in the midst of it all, you have to know how to listen to people.

In the end, if you're going to have a successful parley you have to be patient, respectful, tolerant, intelligent and good at negotiating. Not everyone combines these qualities. There are some further requirements as well that I can't go into here, for they would take a whole story in themselves. To wrap up, let me offer some lessons from experience on how to use the parley.

The parley can teach us some lessons

Everything I have said above points to the fact that the parley, which is a real village institution in Burkina Faso, constitutes a space for dialogue or for participatory development communication. Yet, there are some requirements that must be respected if you are going to use it wisely. You must never forget that it will take a lot of time, and that using the techniques of group dynamics can help to make it run smoothly.

As you can see, the stories I have just recounted offer some real lessons. We saw how the parley allowed the village of Silmiougou to put an end to its water wars. In the eastern villages, it put out the bushfires. That's why we said in the title of this story that the parley can save villages. Now you understand why Grandpa kept talking about how 'the parley saves the village'. Now you can tell everyone that the parley – the African-style town hall meeting – is a practice that Burkina villages, like those elsewhere in Africa, use as a method of PDC. All we have to do is refine it so that it can become a real development tool. For all these reasons, I recommend its use in participatory development communication. That's the end of my story, and it's time to say so long and take care.

Your cousin from the village.

Notes

1   This is a conservation technique that consists of burning the bush immediately after the rainy season and before the vegetation, particularly the grass, has completely dried out in order to encourage regeneration.

2   Cousins à plaisanterie are a Burkina social innovation where different ethnic groups mock one another, which allows people to joke about sensitive subjects.

3   These protected areas are designated natural resource conservation zones, where management includes reforestation and fire prevention.

4   Cola are nuts that contain a stimulant substance and play an important symbolic role in social intercourse.

5   Dolo is millet beer.

V

Collaborative Learning in Participatory Development for Natural Resource Management

Forging Links between Research and Development in the Sahel: The Missing Link

Claude Adandedjan and Amadou Niang

Rural people in the Sahel, where the economy is based largely on farming and forestry, are still using inefficient production technologies. Moreover, the advance of the desert and the collapse of ecosystems that the region has witnessed for several decades are steadily impoverishing the population.

Researchers at the International Centre for Research in Agroforestry (ICRAF) have been working for several years to develop new technologies

that could improve output and thereby raise incomes and reduce poverty among rural people in the Sahel. Some of the most promising innovations have included:

- building fodder stockpiles so that livestock can be better fed, especially during the dry season;
- food banks to improve rural people's nutritional health and to save from extinction certain plant species, such as the baobab, that nourish and protect village fields;
- planting hedges to keep livestock from straying into fields and allowing more intensive planting of crops;
- improved and domesticated cultivars of forest fruit trees to raise orchard output and incomes;
- improved fallowing methods for restoring soil fertility and raising crop yields.

At the outset, ICRAF's strategy was to promote adoption of these innovations by working with development institutions and their agricultural extension services. ICRAF's contribution to this partnership was in the form of training, information and raw materials.

Yet, this strategy was not very successful. A number of reasons were put forward to explain its failure. First, the innovations were not based on the kind of local knowledge and expertise that the communities possessed, so they were little involved. Extension workers did not have the required technical expertise and were chronically short of resources. Moreover, no account was taken of the way in which farmers viewed these innovations. Finally, the extension services showed little interest in promoting the innovations, feeling that their tasks were simply designed to help the researchers.

Consortiums were created to address the situation by strengthening interaction between researchers and development workers, and to reinforce linkages by fostering an inter-agency team spirit.

While this new strategy improved the research–development relationship, rural communities still found themselves excluded from research and development discussions. Moreover, the strategy still failed to take into account farmers' perceptions and their resource and land management strategies.

In the end, despite all the effort that ICRAF has put into popularizing these technologies, it must be admitted that the adoption rate is still very low.

A different approach to research

Since 2000 a move has been under way among development stakeholders in the Sahel to exchange ideas and examine the reasons for these failures. Yet, while efforts initially focused on finding more efficient ways of disseminating the innovations, ICRAF has recently shifted its focus towards more participatory approaches. Field studies have shown that however effective the institutional partnership might be in getting research and development players to work together, when it comes to the farmers themselves the approach has remained very vertical. ICRAF has also recognized that most of the technologies proposed were based on a top-down model – that is, they were conceived and designed by researchers without any input from farmers. The researchers then had to try to 'sell' the innovations to the farmers. Participation was limited to ways of consulting farmers and rallying them around the activities proposed by the researchers.

Given the disappointing results from this approach, it became clear that the problem lay not only in the way in which the innovations were being disseminated, but in the very concept of the research process itself. The discussion sessions showed clearly that if people are not involved in the process from the outset, they will persist in seeing it as irrelevant to their concerns and their needs.

By way of example, a member of the ICRAF team recently related that some time after planting hedges in a community where stray livestock were ravaging crops, she went back to the village for a follow-up visit. To her great surprise, the people greeted her by asking whether she had come back to see 'her' trees. They felt so detached from the experiment that they still looked upon the trees as the property of the person who had planted them.

In another case, a participant in a training session reported that, after travelling several hundred kilometres to plant hedges in a village, he and his colleague were accorded a very chilly reception by the villagers. Some of them stayed away, refusing to come anywhere near the visitors. Taken by surprise, the two extension workers wondered why they were so unwelcome, particularly since they had obtained the consent of the village chief in advance. At this point the chief confronted them in person and angrily demanded to know what they were trying to do to his village. He insisted that they had abused his trust and had come solely to bring disaster on the villagers. Deeply disconcerted, the visitors finally realized that the type of shrubs they had brought with them were seen by the main local ethnic group as portents of evil spirits who would destroy the village.

Many examples of this type can be cited, where needs are analysed and solutions identified in a vacuum without taking into account local systems of understanding and the worldview that underlies them. Naturally enough, ICRAF has turned its thinking towards a new way of doing things. It was clear that it would have to think about more participatory approaches, in the course of which the following questions arose:

- How can we get farmers to participate more actively?
- Do these technologies really take account of people's priority needs?
- Among the target population, are all social groups considered, especially small farmers, women and youth?
- How can we help local people to participate in co-generation and co-dissemination of technologies?
- How can we ensure that our research results are actually put to use?

- Which tools of participatory communication must we use to promote change and innovation in strategies and methods?

Experimenting with a new approach

ICRAF-Sahel and its partners are currently undertaking a wide-ranging programme of capacity-building for stakeholders involved in farming and forestry in the Sahel in order to overcome the inadequacies of the current approach. To start with, three countries have been targeted: Mali, Senegal and Burkina Faso. There are 45 people participating from five different kinds of institutions: research facilities, training centres, development organizations, non-governmental organizations (NGOs) and women's associations.

The main challenge facing this consortium will be to involve rural people effectively in research and development efforts and in this way to launch a process of co-generation and co-dissemination of agroforestry innovations. The new approach will pay particular attention to farmers' knowledge and expertise, and this, in turn, will require a thorough familiarity and understanding of the local context.

Special attention will have to be given to strengthening consortium members' capacities in the use of participatory communication methods and tools, recognizing that the consortium consists for the most part of researchers who are trained in the 'hard sciences', and are consequently too rigid in their attitudes and practices. This effort will, in time, encourage them to change their practices and attitudes so that they can work more effectively with other stakeholders, particularly farmers.

The team is aware that it will have to reach out more effectively to farmers, with whom interaction is currently very weak. In fact, while the consortium has established a working relationship with farming organization representatives, their viewpoints do not necessarily reflect the aspirations of the different groups in rural communities.

Furthermore, the team hopes that, with the collaboration of these stakeholders, it will in time be able to develop and propose solutions and activities that will meet the needs of the various socio-economic groups in rural communities. The agroforestry innovations put forward in the past represented prototypes that took no account of the particular needs of certain groups. This was true, for example, in the case of small farmers, for whom the technology of 'living fences' remained inaccessible because they lacked the additional funds needed to adopt it. By revising their current approach to development communication to make it more participatory, ICRAF and its partners hope to strengthen mutual understanding among stakeholders, and to give new impetus and meaning to their research efforts, which, in the end, are devoted above all to improving living conditions for rural communities.

Isang Bagsak South-East Asia: Towards Institutionalizing a Capacity-Building and Networking Programme in Participatory Development Communication for Natural Resource Management

Maria Celeste H. Cadiz and
Lourdes Margarita A. Caballero

How are the facilitating factors enhanced and the challenges surmounted in institutionalizing an experience-based distance programme aimed at capacity-building and networking in participatory development communication (PDC) in natural resource management?

In essence, this question states the participatory study problem that the College of Development Communication (CDC) at the University of the Philippines, Los Baños, addresses in implementing its own pilot of Isang Bagsak in the South-East Asian region. This chapter reflects on PDC as something more than a community-based approach. It examines PDC at the project management level and partly at the learning programme level, where its principles also apply.

Isang Bagsak South-East Asia

'*Isang Bagsak!*' is a Tagalog expression signalling consensus, agreement or affirmation in a participatory meeting, and Isang Bagsak South-East Asia is this region's learning and networking programme emphasizing the participation of stakeholders in natural resource management (NRM) through PDC processes. The Isang Bagsak programme thus aims to improve communication and participation among researchers, practitioners, communities and other stakeholders in natural resource management and to reinforce the potential of development initiatives in helping communities overcome poverty. CDC is implementing its pilot of the programme, with the aim of institutionalizing it in the region after a 15-month pilot that included two participants from South-East Asia (Viet Nam and Cambodia) and one from Africa (Uganda).

The programme includes an introductory workshop on PDC and on the Isang Bagsak project; local discussions, practical study and the application of PDC in natural resource management; sharing and discussing syntheses of the previous themes at the regional level in a web-based forum; a face-to-face mid-term capacity-building workshop on specific PDC processes and techniques; and a final face-to-face evaluation and planning workshop.

Isang Bagsak South-East Asia began in August 2003. Three teams were carefully selected to participate in the first cycle of the programme: the Forestry Administration of the Ministry of Agriculture, Forestry and Fisheries in Cambodia (representing the government sector); the community-based coastal resource management programme (CBCRM) based in the Hue University of Agriculture and Forestry and the University of Fisheries in central Viet Nam (representing academe); and the Legal Assistance Centre for Indigenous Filipinos, better known as PANLIPI in the Philippines (representing the NGO and people's organization sectors). The introductory workshops have now been completed and the e-forum has been initiated.

Facilitating factors

The facilitating factors in implementing the programme include the following:

- high qualifications of the implementing and facilitating team, who are experts with advanced degrees and practical field experience in development communication, coupled with their institution's long history in development communication education, practice and research;

- collegial and congenial working relationships among team members reinforcing their high morale and making participatory project management possible;
- reasonable flexibility in administrative procedures of the partner agencies, the University of the Philippines at Los Baños Foundation, Inc and the International Development Research Centre (IDRC), respectively;
- high level of support from IDRC;
- sufficient documentation of the pilot phase of the programme, providing a template for the protocols of the current pilot in South-East Asia;
- prior training and experience of the implementing team members in handling distance education courses; and
- a culture of excellence and innovation in (including a high commitment of) the implementing agency.

Challenges

In spite of these facilitating factors, the implementing team has faced numerous challenges, as follows.

Challenge 1: Clearly presenting and reaching a common understanding of the programme cycle, rhythms and mandate among team members, as well as between implementers and learning participants

Part of the programme is an orientation workshop on the concept and processes of PDC and Isang Bagsak for the implementing team. At CDC, the decision was made to implement the programme as a team, rather than to assign one main facilitator employed by the project to run the programme. It was thus important that the different members of the implementing team, most of whom are already experts in development communication, come to an understanding of what the programme is about and how it is run.

The orientation workshop was also seen as an opportunity for CDC to share its own experiences in development communication with other implementers, and potential implementers, of the Isang Bagsak programme in other regions of the globe, with the aim of building a global Isang Bagsak network. Beyond being an in-house workshop, it thus became an international meeting where like-minded individuals from various disciplines and regions of the globe, all interested in uplifting the well-being of people in areas with critical states of natural resources, excitedly shared their experiences and insights as they discussed the PDC process.

Once the excitement brought about by the prospects of being part of a global network of PDC practitioners and scholars had died down, CDC came to grips with implementing the programme in its region. In an in-house forum requested by members of the CDC implementing team after the orientation workshop, the idea of incorporating PDC at the community level prevailed among team members. Yet, at the coordination level, the viewpoint was that Isang Bagsak, being essentially an experiential distance learning programme for PDC implementers, should primarily help to facilitate the learning process. Perhaps there was a strong desire by team members to directly undertake fieldwork, a valid sentiment coming from development communication teachers and researchers.

This led us again to validate the observation on the nature of communication as a process in project team management. Mutual understanding does not necessarily take place instantly; a single utterance of facts and principles does not bring about immediate understanding or acceptance of an idea. This sometimes made us wonder about our credibility as project coordinators among our own team members, a question that recurred whenever team members kept asking what the IDRC programme specialist would say when the need to settle issues related to implementing the programme – implying that our explanations were not sufficient. Arriving at a common understanding on implementing the programme as a team required plenty of meetings and discussion, as well as juggling schedules to find common time to do so in our multifaceted preoccupations as CDC faculty members.

IDRC has certainly assumed the role of a helpful adviser who does not categorically say 'yes' or 'no', but, in the true spirit of participatory management, refers back questions with his own question on why a certain direction was, or will be, taken. He has likewise taken the stance of openness that Isang Bagsak is a learning process and is in the process of continually evolving, as new and different partners implement it.

In itself, the participatory management style and its perspective challenge traditional ideas of project management and leadership, as well as the extent to which programme implementers may have internalized the participatory paradigm beyond conceptual understanding.

The same interpretation that the experiential learning dimension of the programme is itself the project surfaced within two learning teams. One prospective learning team thought that the programme would support new field initiatives; in another team, participation extended to the community level. It took a meeting with the IDRC programme specialist to clarify with members of the implementing team that Isang Bagsak is a learning programme, not a community outreach initiative in natural resource management – such an initiative being the intended participant of the learning programme.

The important lessons we offer in this experience are:

- the need to clarify the programme cycle, not just its content, by being careful in our use of terms, using credible channels and patiently developing mutual understanding on issues related to programme implementation; and
- incorporating the PDC concept and practices within the management of the programme and specific initiatives – an important dimension of participatory development communication.

Challenge 2: Exercising discernment and negotiation skills in selecting the right learning teams beyond clear-cut criteria set for participants
The process of selecting learning teams was in itself a learning process for the CDC. Misinterpretations of the nature of the learning teams' participation and support were largely brought about by the use of the term 'project' in referring to the learning programme. While the criteria for selecting the learning participants and what was expected from them were clearly spelled out, this was no guarantee that these criteria were clearly understood by the learning participants.

The criteria for selecting learning teams were as follows:

- having an ongoing project in community-based natural resource management funded for the following two years;
- sustained connectivity and access to the internet;
- willingness of the team members to learn PDC through experience;
- willingness of project managers to devote at least ten person hours per week to PDC activities, such as monitoring the use of PDC methods at the site; team meetings; studying the Isang Bagsak South-East Asia implementation manual and other resource materials; participation in the e-forum; and participation in workshops and evaluation meetings.

As the implementing team, the CDC team likewise considered the following factors in selecting learning team participants:

- willingness or interest to learn and adopt the PDC process;
- potential influence to advocate PDC for its institutionalization in community-based projects;
- participatory orientation;
- other practical considerations such as location and travel costs required for monitoring, and the prevalence of peace and order in the community site.

In fact, one of two prospective qualified teams was selected on the basis of need and the potential to institutionalize the PDC process within government. It turned out, however, that a month after finalizing negotiations, the selected team lost its internet connection, a requirement for participation in the programme, and sought support from the programme to restore it. Moreover, the supposed team leader was relocated to a provincial post and the new one, apparently not properly oriented by his colleague or supervisor about participation in the programme, initially gave the Isang Bagsak facilitators a cold reception in organizing the introductory workshop. Furthermore, their supervisor, who had endorsed and warmly received the project facilitators during the initial visit, became inaccessible during and after the introductory workshop, when the new team leader suddenly requested support for internet connections.

This particular experience was perhaps incidental. Yet, one lesson is perhaps that referrals should be an important input, in addition to the information gathered from application papers and visits, with face-to-face interviews when selecting learning participants. Just as students apply for admission to graduate studies, learning teams should apply to the programme and clearly understand the requirements and implications of such participation. Signing a commitment by learning teams and Isang Bagsak facilitators alike may be a necessary step in negotiating learning participation.

On the other hand, unexpected developments such as these continue to underscore that the programme cannot have a fixed system and procedure. Perhaps the more valuable lesson in this experience is that facilitators of such programmes should always be ready to engage in a balancing act between the need to be consistent with certain guidelines and procedures, and the need to be flexible where the unexpected arises.

Challenge 3: Negotiating a realistic programme schedule that allows for optimum participation of four different teams, including the implementing team, each with their own calendar of activities

Negotiating a workable timetable for the programme was in itself a complicated communication process. What made it doubly difficult was the mode by which this was negotiated, by email and/or websites. It seems that we Asians have not yet mastered communication via these channels. Often, a message sent is left hanging, with replies sent back by some recipients, while others neglect to answer even a simple 'Yes, the schedule is alright with us' or 'No, it is not workable; here is a counter-proposal.' Then again, the problem may be the recipients' limited or intermittent access to the internet.

The reality is that participants are busy themselves, and responses to questions about schedules do not necessarily entail a simple 'yes' or 'no'. Often, one's own team members need to be consulted first. Sometimes, due to the length and complicated nature of the process, the act of sending a reply is forgotten altogether.

This also has implications for the time it takes for exchanges on the web and among team members to complete loops in the e-forum. Based on the timetable drawn up, the whole programme from inception to the final evaluation workshop should take about 15 months, taking into account the slack expected during holidays and important occasions. We have received feedback that the time allotted for reacting to everyone's postings – one week – is too short for these series of exchanges within a team, and that two weeks would be more realistic. This would stretch the programme to about nine more weeks, or a total of approximately 17 months.

During the mid-term workshop when members of the three learning teams met together, participants suggested a more flexible schedule for the e-forum: two or three related themes to be simultaneously launched in separate conferences, and learning participants to be allowed to take as long as two to three months to complete the cycle of discussions, the posting of team discussions and debate. The programme is divided into four parts, comprising a total of nine themes. Each part includes two to three themes. The proposal, which has now been adopted for the rest of the e-forum and for the second programme cycle, is for each part with its two- to three-component themes to be launched together, instead of having a linear sequence of all nine themes. This was because the themes within the four parts were interrelated; the logic, therefore, was to discuss them together. However, postings of discussions will be grouped separately, based on each individual theme conference.

Challenge 4: Transcending and accommodating language and cultural barriers, and exercising cultural sensitivity in facilitating the learning process across cultures

In South-East Asia, the medium of exchange on PDC at the regional forum is English, a second language for all the learning teams. Proficiency in English varies across and within teams. Therefore, the process of translating postings and exchanges from the vernacular into English, and the other way around, requires additional time. However, it must be noted that, in hindsight, participants from the pilot phase found one unexpected gain from their participation in the programme: an enhanced proficiency and confidence in communicating in English.

Aside from the language barrier, facilitators found it a challenge to understand the participants' cultural differences and to exercise sensitivity in facilitating learning. For instance, during the introductory workshops, the Cambodian participants took a long while discussing and arriving at a consensus on their collective replies to questions in their own language before they translated them into English. The lengthy discussion is part of the Asian concern for 'saving face', interpreted as a form of social grace by some anthropologists. Asians would not want to belabour or burden the facilitators, who are visitors, with the details of their disagreements, uncertainties and tentative stances. They would just translate for the latter's consumption the resolutions agreed upon, already processed and deemed presentable.

In the e-forum, the lessons posted by participants looked too sanitized and 'correct', thereby lacking in richness precisely because the disagreements, uncertainties and questions were left out. The recently created private conferences among learning team members is an answer to this need for private conversations among members of a learning team before they post their own lessons, fit for 'public consumption'.

Yet, the question remains: how do we strike a balance between the Asian concern for saving face and the learning that comes from making mistakes and acknowledging them? Perhaps, in the Asian context, Isang Bagsak facilitators need to point out the reality and value of mistakes in the process of learning and capacity-building, rather than the view that making mistakes is a stigma. Should these 'mistakes' be kept concealed in private conferences? Likewise, will the habit of avoiding conflict and controversy as part of the Asian/Buddhist ethos of smooth interpersonal relations be a hindrance to participatory learning?

Another cultural challenge to learning PDC are the hierarchical structures and relationships found in many organizations in South-East Asia. One observation in capacity-building programmes for such organizations is that team leaders, who are supervisors, do not take part in the programme activities and do not mingle, but primarily relegate the learning to their team members. On the other hand, the leaders' participation in the capacity-building programme is crucial to ensuring the application of the approach in their national resource management research and activities. Thus, another challenge with regard to capacity-building in PDC for natural resource management is how an organization can learn to apply PDC without contradicting itself in its project management style. Will its credibility in applying PDC be compromised if it does not adopt a similarly participatory project management approach? Or will learning and applying PDC in communities bring about a shift in a hierarchical style of project management, albeit slowly and incrementally, towards a more participatory approach in project management?

On the other hand, inasmuch as our bias is for participation, to what extent does advocating PDC impose on the learning team's non-participatory, hierarchical culture an alien perspective and method that might be inappropriate?

The lawyer/executive director of PANLIPI, however, reminded us of the rights-based perspective that is compatible with the participatory perspective: a hierarchical leadership style may also be consultative and participatory if it upholds the basic human right of constituents to express themselves and communicate their honest views and insights. Upholding people's participation in development, therefore, should not be viewed as an imposition if it upholds basic human communication rights.

Challenge 5: Making the learning process participant driven and experience based, yet striking a balance with expert or theoretical knowledge in the discipline

Aside from its application in NRM communities and its implication in NRM project management, PDC is a philosophy that also applies to the capacity-building or learning process in the programme. Thus, the challenge that facilitators face deals with striking the perfect balance in combining participants' experiential learning with expert or theoretical knowledge in PDC. What is the best way of drawing out experiences and reflective thinking while identifying lessons and insights in participatory development communication? How should facilitators incorporate the wealth of knowledge and wisdom from experts and the existing body of knowledge within PDC when applied to natural resource management? Furthermore, how can a participatory learning process in PDC contribute to further building this body of knowledge?

Challenge 6: Learning as a community/team by alternating local face-to-face meetings in the local language with regional (international) virtual meetings in a foreign language using the internet

Aside from our previous experience, research and practical and theoretical knowledge on PDC, as well as our previous training and familiarity with distance learning mechanics, our only preparation in carrying out the Isang Bagsak programme in South-East Asia was, perhaps, our openness to learn the cycle and rhythms of the programme. We were learning along the way. The newer dimension of facilitating is the community or team mode of learning compared to individual-based distance learning.

The main lesson here is the need to painstakingly spell out this learning cycle and to explain it in detail, making such explanations and exercises part of the programme's orientation. Such exercises should not just focus on using the electronic forum's software, but on the whole process of discussions within a team, synthesizing the discussions, and then posting in and downloading from the electronic forum. This makes it imperative that facilitators are highly familiar with the Isang Bagsak process. Thus, facilitators should have gone through the programme themselves if they are to facilitate learning. A manual on Isang Bagsak facilitation may likewise prove helpful.

Challenge 7: Capturing the collective learning of participant teams while remaining concerned with the relevance of PDC and NRM efforts to enhance the well-being of grassroots communities with and for whom the teams are working

This final challenge spells out the need to strike a balance between the capacity-building process of NRM workers and the benefits of their efforts within the natural resource management community. This is primarily a reminder or caveat to Isang Bagsak facilitators that, in our concern with learning the PDC process, we should not lose sight of 'the big picture': the impacts on the natural resource management community. As the programme continues to unfold, we anticipate further insights and challenges related to the impact of the programme on the well-being of the communities with whom learning participants work. Participatory development communication is, beyond a body of knowledge and a practice, people's lives, as the PANLIPI executive director reminds us.

In all these challenges, the College of Development Communication takes comfort in the shared view with its regional partners of Isang Bagsak as an evolving programme and network that is continuously redefining itself. Its dream is for the Isang Bagsak programme to eventually evolve into a regular self-sustaining certificate distance-learning programme in PDC in natural resource management and other development concerns for various types of development workers in different contexts.

Implementing Isang Bagsak in East and Southern Africa

Chris Kamlongera and Jones Kaumba

This chapter seeks to show how the Southern Africa Development Community Centre of Communication for Development (SADC-CCD) is assisting national bodies working on environmental and natural resource management (NRM) in Malawi, Uganda and Zimbabwe. With support from the International Development Research Centre (IDRC), SADC-CCD is working on a project aimed at building the capacity of institutions

in participatory development communication (PDC). This work is part of ongoing activities at the centre, aimed at assisting governments of the region in their development efforts.

The problem with these efforts has been their failure to put ordinary people in the driving seat. The participation of rural communities (who constitute more than 70 per cent of the total population of the region) has never really been seriously considered as critical to bringing about positive change in the region. The voices of the people have not been seen as important in most development efforts. These voices are often, if not always, ignored by those who make decisions on development issues.

Those who agree that people's participation in rural development and poverty alleviation efforts is critical are still grappling with how best this can be brought about in a sustainable manner. There is still a need for a body of practical knowledge on how to involve rural communities fully in such work. SADC-CCD was set up to find ways of doing this.

To date, it has come up with participatory communication methodologies that are trying to answer the quest for such a body of knowledge, such as participatory needs assessment for development communication and rural development; participatory communication strategy development; participatory curriculum development for farmer field schools; participatory rural communication appraisal; including rural communities in the writing of proposals; and participatory evaluation of communication programmes and the use of folk media.

These participatory communication methodologies have been tested across several issues in rural development and have been seen to work. What remains as a challenge to the SADC-CCD is sharing the body of knowledge so far generated in a manner that is both cost-effective and sustainable.

Addressing the problems of rural development

Low participation of people in poverty reduction and rural development programmes often results in a low utilization or failure of these programmes. Explanations put forth to explain these shortcomings include poor planning with communities to be involved; a low sense of ownership by communities; inappropriate technical solutions; poorly packaged information and knowledge; ineffective training methodologies for rural/semi-literate clientele; and ineffective communication channels and wrong target groups. In many cases, these problems can be addressed by Communication for Development, a cross-cutting approach applicable to any area of development where lasting progress depends upon the informed choices and actions of the people involved. It applies equally to programmes for improved agriculture; nutrition; food security; health; water and sanitation; gender awareness; population and reproductive health; livestock; forestry; the environment; literacy; rural credit; income generation; and other key areas.

Communication for Development consists of the systematic use of communication to effectively involve people in development, most particularly in rural development. It is based on the principle of dialogue, using communication approaches, participatory activities, media and channels with all levels of people concerned as equal partners Instead of 'target groups' that characterize Western-style advertising and promotion, Communication for Development is mostly based on 'interaction groups', promoting a common understanding and exchange of knowledge and experience. It can be used at the interpersonal, group and mass levels.

Communication for Development can:

- ensure that planning takes into account the community's underlying concerns;
- build community ownership of projects and empowerment;
- ensure that the best technical solutions are considered, taking into account community perceptions and practices, as well as their indigenous knowledge;
- package information in ways that are useful and attractive to users, particularly rural people who may be semi-literate and/or traditional;
- enhance training and technology skills transfer with communication approaches and media materials suited to people with little or no formal education skills;
- identify both effective and preferred communication channels and media, influential sources of information and advice, and modern and traditional knowledge; and
- help to safeguard against biases and bring out the concerns of beneficiaries who may be marginalized.

Communication for Development can produce both qualitative and quantitative results, measurable as changes in awareness, knowledge, attitude and practices. Moreover, it is particularly geared for use in rural areas, with more traditional cultures, although its principles may be applied equally in the peri-urban context.

In order to address the problem of the lack of participation in rural development programmes, SADC-CCD has developed and used a methodology comprised of the following seven phases:

1 situation assessment;
2 participatory research with the community: problem and solution identification; baseline survey for awareness; knowledge, attitude and practice surrounding a specific development issue (a baseline survey whose instrument is built on the results of the participatory research also provides a benchmark for future evaluation of the communication programme);
3 communication strategy design;
4 participatory design of messages and discussion themes;
5 development of communication media materials and methods to be used;
6 field implementation (with training of field staff as necessary); and

7   monitoring and evaluation, including second survey (to measure results and plan the next intervention, if necessary).

This methodology has been adapted, tested, marketed and disseminated. The solution has been based on methodologies that include participatory approaches which aim at actively involving people, at all levels, in identifying rural development problems and solutions, sharing knowledge, changing attitudes and behaviours, making decisions and reaching consensus for action.

It has been disseminated through experiential workshops, advisory and consultancy services, and the production of communication materials using participatory methods. Alongside the methodologies cited earlier, the centre is producing manuals, case studies and handbooks to go with the workshops.

Dissemination of the methodology itself has been a key factor leading to its adoption by rural development programmes and communities. A strategy for inducting national staff teams from rural development projects into innovative 'learn-while-doing' and action-oriented workshops that include fieldwork and follow-up implementation at the grassroots level has been developed.

This work has produced noticeable and positive changes in the awareness, knowledge, attitude and practice of those involved, including rural communities, development field staff and their employers in government, non-governmental organizations (NGOs), institutions and international organizations.

In pursuit of its new strategy of partnerships, SADC-CCD has been working with some organizations with which it shares common understanding of the importance of Communication for Development. For example, it has been working with the Centre for Rural Radio Development (CIERRO) in Burkina Faso in an attempt to improve the status of rural radio in Africa. It is also working with IDRC on PDC in environmental and natural resource management. This latter programme operates in Uganda, Malawi and Zimbabwe as the Isang Bagsak East and Southern Africa (IB-ESA) programme. The programme provides one concrete example of participatory development communication at work. Let us look at how this is taking place.

Isang Bagsak in East and Southern Africa

The introduction of the PDC programme for natural resource management in East and Southern Africa has been welcomed as a major step in efforts aimed at improving the management of the environment, natural resource research and development initiatives in the region.

The programme recognizes that the level of communication skills of those attempting to implement development programmes and projects is vitally important to the success of any efforts whose objective is to ensure sustainability. In the main, the programme is expected to help participants design and apply new ways of reaching people more effectively, wherever they may be, through interpersonal, group and mass communication. It also aims at making the policy development process more transparent and open to all stakeholders and involving people in decision-making processes of designing and planning possible solutions to problems in their society. Furthermore, it seeks to help participants design multimedia development messages, materials strategies and campaigns to communicate new ideas and practices to those in need of them. Finally, it strives to assist governments through their various institutions and NGOs to formulate well-researched policies for reducing poverty and ensuring sustainable, gender-balanced and environmentally friendly rural development.

By improving the ability of researchers and development agents in working with communities, Isang Bagsak will open up whole new possibilities for communities to handle natural resource management and development issues in their own way and in line with their traditional and cultural requirements, thereby contributing to the sustainable use of available resources.

Recognizing that capacity-building in PDC should be an important programme area in the field of NRM, IB-ESA – through its eight projects spread out across the East and Southern African region – is striving to make a difference. Researchers see it as an important vehicle for working with communities.

Thus, the Isang Bagsak programme is providing face-to-face interaction and tuition between participants and facilitators, as well as interaction through an electronic forum. Participants can also communicate among themselves freely through a computer-based 'Village Square'. Moreover, they have access to the IDRC library resources, as well as SADC-CCD materials and books.

The programme has also ensured that these human and materials resources help participants to build their capacities to work with local communities in a participatory way, thereby making it easier to develop partnerships with other development stakeholders in their quest to influence effectively the policy environment at local and national levels. It has also offered participants an opportunity to quickly obtain up-to-date information on participatory methods in natural resource management without having to leave their place of work to receive such training. Most importantly, Isang Bagsak East and Southern Africa has promoted South–South cooperation to ensure that participants in the South draw on each other's expertise and experience in order to reduce dependence on the North. It is also noteworthy that participants in this initiative need not have the same level of experience since teams with little or no experience can share and learn from their more skilled counterparts.

The initiatives enrolled in the programme come from a variety of disciplines, which makes the implementation of PDC a multidisciplinary one. Today, three countries are participating in the programme: Malawi, Zimbabwe and Uganda. As described in the following sections, the supported initiatives greatly differ from one another. Yet, they all share the same views regarding the need to empower local communities and to increase their sense of ownership of research or development initiatives.

Malawi

Three initiatives are currently under way in Malawi. The first one, entitled the Macadamia Smallholder Development Project, is being implemented by the Department of Agricultural Extension Services (DAES), which is part of the Ministry of Agriculture, Irrigation and Food Security. Since 2000, DAES has a new policy that advocates pluralistic and demand-driven extension services. The challenge is to empower farmers so that they are able to demand services which address their needs and problems. At the same time, the department has to coordinate the activities of various extension service providers operating in the communities.

The Macadamia Smallholder Development Project covers two agricultural development divisions – namely, Kasungu and Mzuzu in northern Malawi. Both researchers and agricultural extension staff participate in implementing its activities. The initiative aims at improving the well-being of Malawians through poverty alleviation among rural people by promoting agricultural development. More specifically, it seeks to promote the production of macadamia nuts intercropped with other food crops, thereby ensuring food security and cash income for rural farmers. Conservation of the environment is an integral part of this initiative. According to the team leader, through their participation in Isang Bagsak, the team intends to acquire effective communication skills that will enable frontline extension staff to engage in dialogue and create a mutual learning environment within the communities with whom they work, thereby empowering farmers.

A second initiative, which focuses on indigenous fruit trees, is being undertaken by the University of Malawi's Department of Chemistry. Previous studies on the chemistry of indigenous edible wild fruits growing in Malawi have revealed their great nutritional value, including vitamins A, B, C and minerals. This initiative will focus on the development of those products and the promotion of processing and marketing at household levels in rural Malawi. More specifically, the research will address the utilization and commercialization of indigenous fruits of the Miombo eco-zone. Implementing this study requires working with communities, especially with women, who are the main processors of food in Malawi. Technologies that can remove the drudgery and labour load among the women processors have been developed. Further work in capacity-building of rural food processors to increase value and income for improved household welfare will also be undertaken.

The team also facilitated the identification of priority research and development activities among rural communities. These activities have necessitated working with NGOs and government departments, especially research and extension services.

Finally, a third initiative spearheaded by the Department of Biology of the University of Malawi is also under way, which has been involved, in the past, in various multidisciplinary projects where participatory development communication was lacking. In this case, the initiative will focus on water resource management in southern Malawi.

One of the most interesting aspects of this initiative is that it stems from natural scientists' interest to incorporate social issues within natural resource management research as a way of improving their work. So far, team members have carried out preliminary data collection. The five-people team will soon be conducting detailed surveys where questionnaires and focus groups discussions will be used to collect data. They believe that PDC can greatly help in this process. Among other things, it is believed that PDC will contribute to increasing the indigenous and modern community knowledge base by collating indigenous knowledge that already exists at the local level and using it as a stepping stone towards the development of water resource management plans. Another objective of this initiative consists of assessing the microbial and chemical quality of water to determine if it is suitable for human consumption. The relevant gender issues affecting water resource management in the communities in the study area will also be examined. Finally, this research will also identify the factors that contribute to the degradation of the catchment area and will propose mitigation measures that can be put in place.

This increase in knowledge, however, can only be shared and utilized effectively if the team members are equipped with appropriate skills through the PDC training programme.

Uganda

Three initiatives are also taking place in Uganda. The first, entitled Integration of Natural Resource Management in National Agricultural Advisory Services (NAADS), is located in Kabale, in the sub-county of Rubaya.

The area is mountainous and very steep. Farmers continually face a problem of soil erosion as they farm on steep slopes. Currently, the programme is encouraging them to integrate the planting of trees, such as apple trees and cariandra, within their farming in order to control soil erosion.

Interacting successfully with the communities involved in this initiative is vital to its success, and PDC is seen as the best methodology to do so. According to the programme's coordinator, this new methodology, which ensures the participation of the communities at every stage, is probably the best possible solution to tackling some of the communication issues inherent to the projects.

The second Ugandan initiative is being implemented under the responsibility of the Kawanda Agricultural Research Institute (KARI) of the National Agricultural Research Organization (NARO). It deals with communication among banana growers regarding soil and water management, post-harvest handling and the improvement of cropping systems.

The communication issues being addressed by the project include the need to share farmers' indigenous knowledge, the inadequacy of extension work due to poor facilitation and motivation, farmers' lack of access to adequate information, as well as the existence of communication gaps between farmers and researchers, farmers and farmers, and researchers and extension staff.

Farmer-to-farmer communication through posters, brochures and open days has been undertaken in order to alleviate the problem of inadequate extension staff. Joint planning between researchers, farmers and other participants has been contributing to the reduction of communication gaps among the different stakeholders.

Finally, the third initiative undertaken in Uganda deals with sustainable land use in banana production in central Uganda. The main partner for this initiative is an organization called Volunteer Efforts for Development Concerns. This initiative aims to conduct participatory rural appraisal sessions to develop food security calendars and action plans; train rural development extension workers in farm planning and layout, farming as a business, sustainable agriculture, post-harvest handling and communication; establish on-farm demonstrations; provide quality planting material (high yielding, pest- and disease-free); and conduct agricultural practical training in relation to spacing, organic manure preparation, integrated pest management, pruning/de-suckering and post-harvest handling.

Since this work involves substantial interaction with farmers, it is vital that a suitable methodology is used to ensure optimum results. Participatory development communication seems particularly appropriate in this case.

Zimbabwe

In Zimbabwe, an initiative called the Desert Margins Programme is being implemented with the aim of arresting land degradation in the desert margins through demonstration and capacity-building activities. The programme addresses issues of global environmental importance, national economic and environmental concerns and, in particular, the loss of biological diversity, reduced sequestration of carbon and increased soil erosion and sedimentation.

The initiative is part of a larger programme covering nine countries: Burkina Faso, Botswana, Kenya, Mali, Namibia, Niger, Senegal, South Africa and Zimbabwe.

Key sites harbouring globally significant ecosystems and threatened biodiversity have been selected in each of the nine countries. These sites are to serve as field labs for demonstration activities regarding the monitoring and evaluation of biodiversity status, the testing of the most promising natural resources options, as well as the development of sustainable alternative livelihoods and policy guidelines.

The approach is intended to be a holistic one, as the Desert Margins Programme takes an innovative participatory and integrated natural resource management approach that consists of the conservation of biological resources through restoration activities that reverse degradation processes, rather than the preservation of specific ecosystems or species in protected areas.

In Zimbabwe, the chosen sites are Matobo, Chivi and Tsholotsho, which are based on communal rule. This brings in issues of consensus-building before any work can be done. Ownership issues, policing systems, who controls resource utilization, who is going to implement the interventions and monitor them, who says there is a problem are all issues that must be discussed from the onset.

PDC has a role to play right from problem identification, including finding and choosing possible solutions. So far, work has involved surveys to gather baseline data on the areas. Some tools have already been used – for example, historical timelines and trend lines. The initiative will last six years and the intention is for the initiatives to be continued beyond that.

In this case, PDC comes in handy in empowering the community, thus facilitating the sustainable use of natural resources. Communities will be exposed to an open platform to discuss their natural resource management practices, problems, needs, opportunities and solutions. As asserted by the team leader, 'They will, in the process, gain knowledge to implement their initiatives through guidance from the facilitators.' PDC, therefore, becomes an important tool that ensures ownership of the project by the communities themselves, thereby giving it a much higher chance of success and resulting in improved livelihoods.

The Sedgwick Agricultural Development Project is located in the Tsholotsho district of Matabeleland North Province. It operates on 10,126ha of the estate, which is bound by a number of newly resettled farming house-holds, where it works directly with 84 households and extends services to four other neighbouring villages from the Tsholotsho communal areas.

The project aims to ensure strategic crop production for national food security through irrigation development initiatives. It also seeks to develop capacity for the production of grain crops and other cash crops by the community and to enhance its livestock management practices, particularly animal nutrition and disease control. Finally, it aims to provide breeding services in order to preserve the indigenous Nkone herd for the community and to establish veld management systems, in collaboration with the Department of Livestock Development and Production.

Participation in the project was initially through the local authorities and leadership. The team had to arrange meetings with members of the community within the project area. Areas of intervention were identified by conducting problem tree analysis and ranking through participatory approaches.

This helped the team to ensure that the causes and effects of the problems, as well as their solutions, were identified from the community's point of view. Overall, people felt that the area is a drought-prone one, so there are bound to be problems with water and pastures. These problems, in turn, lead to animal deaths, ill health, low weight and, eventually, the loss of income from the sale of animals. Following the success of this new approach, the farmers have agreed to take part in the development of grazing management schemes and to coordinate the provision of water for dip tanks.

Challenges

Since the beginning of the Isang Bagsak programme, there have been some experiences or some questions that require addressing. These are:

- The lack of backstopping activities: these would ensure that researchers are assisted on the spot while in the field and would allow for effective documentation of activities by coordinating institutions at the field level.
- Accreditation: although the SADC-CCD gives certificates of completion to its training and workshop participants, the participants' organizations or institutions do not recognize these certificates. Recognition or accreditation coming from an academic institution would not only be recognized by other institutions, but would go a long way in rewarding genuine participation in the Isang Bagsak programme and could be used or added to in a degree programme.

Observations and reflections

The PDC programme has so far demonstrated a positive demand for its services to a diversity of stakeholders, as evidenced by the organizations that are currently participating. Bottlenecks must still be removed, however, and the programme must be offered to more research and development institutions in the region.

So far, the sharing of experiences through the e-forum is proving to be an innovative way of getting over the usual hurdles of embarrassment, fear and reluctance to open up that are often associated with adult learning. The facilitator seems to disappear from the scene once the 'theme for the day' has been introduced. The participants then comment on the facilitator's initial statements. They do this freely, digging deep into their past experiences and observations of life around them (including past mistakes) without any fear of embarrassment.

Participants also learn from each other as they post their own responses on the forum. Very often, they respond to their colleagues' entries, as well, seek advice or make comments on their own experiences or problems.

One very gratifying result of the Isang Bagsak experience is the readiness, by scientists, to consider issues of communication. This is a major breakthrough for the programme. Some of the academic participants are already changing their approach to hardcore science teaching, as the Isang Bagsak experience is asking them to consider how, in reality, they relate to the world around them. In doing this, hardcore science is being 'humanized'.

Reflections on Participatory Development and Related Capacity-Building Needs in Egypt and the Arab Region

Waad El Hadidy

Non-profit and other civil society organizations in the Arab region now face difficult challenges. On the legal front, the battle for redefining the boundaries of the non-profit sector is taking place. On the political front, the question of what participation means and how it can be controlled is intensely debated. On the social front, disruptions are affecting longstanding strictly codified value systems and social structures. All these challenges are questioning current development practices. Although both government and non-governmental organizations (NGOs) realize the limitations of top-down, off-the-shelf interventions and sense the need for participation, participatory development has not become deeply embedded in practice.

What can we learn from a decade of grappling with the concept of participatory development and its application? This chapter presents the Centre for Development Services' thoughts on participatory development in the Arab region, with particular insights from Egypt. Participatory development here refers to both the paradigm and the participatory approaches that make it possible, such as participatory development communication (PDC), participatory rapid appraisal (PRA) and participatory learning and action (PLA). Related capacity-building needs will also be discussed.

Participation as a societal value

Many Arab societies were pioneers in ingraining the value of participation in their lives. Egypt, for example, was the first Arab, Muslim and African country to experience modern civil society organizations, as early as 1821. It was also the first to flirt with democratic governance, starting in 1866 (Ibrahim et al, 1991) The first modern NGO, the Egyptian Hellenic Philanthropic Association, was established in Alexandria in 1821. This association was qualitatively different from earlier traditional religious endowments, which were a function of single charitable individuals or families, a form known in Egypt for centuries before.

From a communal-cultural perspective, the diverse reality of civil society in the Arab world, which includes informal social networks and traditional organizations of kinship, tribe, village and religious community, is a reflection of a wide range of civic behaviour that honours participation (Saber, 2002). In Kuwait, the *diwaniyya* is a place where men and, during recent years, women meet informally and is widely acknowledged as the place where the current move towards democracy in Kuwait began (Al Sayyid, 1993). In parts of Lebanon, the tribal system creates a space for consultation among village members over issues that affect their community.

From a religious perspective, participation has always been a core pillar of Islam, especially in non-secular regimes, which have historically prevailed. As a concept and as a principle, Shura in Islam does not differ from democracy. Both Shura and democracy arise from the central consideration that collective deliberation is more likely to lead to a fair and sound result for the social good than individual preference. Both concepts also assume that majority judgement tends to be more comprehensive and accurate than minority judgement. As principles, Shura and democracy proceed from the core idea that all people are equal in rights and responsibilities. The Qur'an mentions Shura as a principle governing the public life of the society of the faithful, rather than a specifically ordained system of governance. As such, the more any system constitutionally, institutionally and practically fulfils the principle of Shura or, for that matter, the democratic principle, the more Islamic that system becomes (Sulaiman, 1998).

The paradigm of participatory development

Despite such deep-seated participatory principles in culture and religion, the paradigm and the institution of participatory development are not so well entrenched in the Arab world. Many international organizations have been keen on engaging local communities in decision-making regarding development initiatives in Egypt. Much of their earlier work during the 1970s began with working with committees of local citizens, as mobilized by these organizations. In order for the work of these committees to continue endogenously, they were advised to form NGOs. This was the formula pursued by most foreign organizations operating in Egypt during the 1970s and 1980s. Ironically, and despite their efforts, foreign organizations were met with theories of conspiracy and doubts about their real intentions.

For example, in a village of the Beni Souef area where an international organization had begun working, members of the committee were particularly doubtful. Throughout the initial meetings, the group of leaders requested a statement of expenses in order to better assess the situation. They found out that US\$160,000 had been spent in meetings and trips between Brussels and Cairo. 'We didn't understand development at the time, and thought that every penny should be spent on improving conditions', stated Mr Badr. As members of the local council, they decided to expel the institute from Beni Souef. But the local leaders were pressured by the US embassy and the governor to let the organization stay. They were advised that 'This organization is going to stay, whether you like it or not. So it would be wise of you to make the best of them.'

While the essence of participatory development had been practised since the 1970s, it wasn't until the 1990s that the written rhetoric began to spread. Training courses and development projects all became prefixed with the term 'participatory'. Some organizations, including the Centre for Development Services and the Near East Foundation, took on the task of adapting the concept to the Arab region. Naturally, the efforts of a few organizations were not enough to reach thousands. Although many organizations are now using the language of 'participatory development', the conceptual framework is not well internalized since few genuine participatory development initiatives are taking place on the ground. Perhaps this is because 'participation in development' is often perceived as a Northern agenda, despite the fact that it first emerged in South America through the writings of Paulo Freire (1970).

There are limited known examples of successful participatory development initiatives. The documentation process of such initiatives usually targets the donor agency rather than local practitioners, decision-makers or community members. The critical reflection processes that precede documentation for the latter groups and result in organizational learning are frail in the Arab region. On the individual level, practitioners seldom engage in critical reflection because the predominant education systems do not encourage it. Organizational cultures are also non-conducive with regard to these aspects. Knowledge generation in our region usually revolves around an urgent need, whether it is a problem requiring resolution or an issue in need of addressing. The regular and consistent communication processes prevalent in rural communities, such as the tribal councils in Jordan or the Qat gatherings[1] in Yemen, create space for discussions, for local wisdom to emerge and for learning to take place. Such processes are not normalized within development organizations or among practitioners. This situation restrains contemplation on both an individual and group level and perpetuates the existing lack of knowledge about participatory development.

Moreover, the policy environment does not always enable civil society to apply participatory approaches. While it is true that in Egypt, Morocco, Tunisia and Algeria, governments encourage NGOs to complement public spending on social services, they fail to allow these organizations to freely empower people for fear that things may get out of hand (Kanawati, 1997).

Reflection on the Centre for Development Services' attempts to turn 'participatory development' into a meaningful concept in Egypt provides interesting insights. Indeed, using the PRA methodology adequately made a significant mark since many NGOs talked about participation as an integral part of their work. However, the emphasis on the NGO sector unnecessarily reinforced the separation between the government and NGO sectors. The overwhelming participatory discourse and its focus on NGOs implied that government cannot be participatory, and that the role of NGOs is to correct for the mistakes that government makes in planning development without involving people. The dichotomies created by the use of words such as 'putting the last first' and 'the marginalized', which dominated the discourse, tended to pit NGOs against government. In such a polarized situation, participation advocates were faced with what they wanted to avoid: a rift between NGOs and government. This only shows that even when efforts in promoting participatory approaches are expended, unwanted outcomes may result.

One region, many peculiarities

When speaking about the Arab region, it is important to realize the differing social, political and economic contexts that enable or stifle participatory development across different Arab countries. For instance, Egypt's social context is one of a patriarchal and authoritarian society reflected in communities' attitudes of dependence and ambivalence. For a long time, the government has taken on the role of the benevolent dictator invoking an attitude of 'government knows best'. The work of NGOs is largely controlled through legislation and scrutinized by state security. As a matter of fact, a study conducted to assess grassroots participation among Egyptian NGOs through indicators of membership, frequency of general assembly meetings and proportion of voluntary to salaried staff concluded that overall participation was very low, due in part to government inhibition of participation (Ibrahim,

1996). The situation in Palestine can be described as the reverse. In the absence of government, NGOs have taken on the role of provider of services, which in other countries would be categorized as public services. NGOs are dominant in Palestine, and the Palestinian Authority is now trying to position itself within the development realm. Such variations suggest that participatory development takes on different meanings, shaped by each country's context.

Implications for capacity-building

There is no doubt that more efforts are needed to contextualize participatory development so that it is more than a prefix to NGO programme titles. The question is how should this capacity-building take place?

The Centre for Development Services was originally established as a resource and service centre for NGOs. Training and technical assistance used to comprise most of its portfolio before it delved into long-term, self-conceived development programmes. Reflecting on the training-and-technical-assistance days led us to realize that this 'delivery of resources' mode of operation in the form of transfer of know-how and skills was essential but not sufficient. It implied that the beneficiaries of capacity-building require the transfer of resources from those who have to those who have not.

This has been and still continues to be the paradigm for capacity-building as perpetuated by NGOs and international agencies alike. As excerpted from a World Bank working paper:

> *... respect for independent civic action does not mean leaving community groups on their own to implement initiatives by trial and error. They generally lack specialized knowledge and the ability to apply it, and the success of their endeavours often hinges on receiving appropriate and sustained technical assistance in fields such as management information and project control, human resource development, and project formulation, monitoring and evaluation. (Siri, 2002)*

It is rather paradoxical for organizations who aim to promote participatory development to be implementing capacity-building in such a way. Being participatory does not only mean engaging people in a transfer of knowledge, but rather shifting the outlook of capacity-building in order to recognize that people already have skills and abilities that only need to surface. This new outlook also recognizes that people internalize new knowledge not only through the transfer of knowledge, but also through experiencing this knowledge themselves.

More specifically, taking the example of participatory development communication and the Centre for Development Services' minimal, yet insightful, experience in training Arab organizations and providing technical assistance to a team of researchers on the concept, it would not be unfair to say that the essence of PDC was not conveyed. Discussions with practitioners who attended the training showed that their understanding of participatory development communication was limited to the production of materials. The research team perceived PDC as an extractive tool used to facilitate information gathering in research.

However, PDC, like other members of the participatory development family, is rather fluid and reinforces an alternative form of communication and partnership between communities and development practitioners. It is about communication, a process integral to our lives. It is also about being cognizant of communication needs, such as facilitating articulation, enabling collaboration and providing understanding, in addition to development needs that may focus on more concrete issues. Unlike capacity-building, which requires a 'how-to' approach, such as proposal writing or business planning, capacity-building in PDC should focus on recognizing that communication is a natural process. Participatory development communication supports the development project cycle. Hence, it addresses practitioners who are engaged in development. Rather than 'teaching' them about participation, it is important to recognize that these practitioners have rich experiences and what is needed is to facilitate their own learning. Rather than 'providing resources', an approach is required for 'facilitation of resourcefulness', a process of learning to learn. It is also critical to realize that participatory principles are not alien to the Arab world's cultural and religious heritage, but have deep-rooted traditions that need reviving.

For Arab practitioners, space is needed to reflect on experience and to develop insights, document stories and share with others. In a workshop held in Jordan in 1997, participants from nine countries in the Middle East reflected on the lack of documentation of local participatory practice, and indicated their enthusiasm for a forum that brings them together to discuss their experiences in the field (NEF and CDS, 1997). Before now, some attempts had been made; but none were sustained, probably because discussions were too formal and abstract and ended up as a series of sporadic and detached topics. Motivation factors were not integrally considered.

The suggested strategy for capacity-building assumes that self-reflection generates knowledge that is profound and better internalized than knowledge imposed from the outside. After all, how can we presume to intervene in others' development if we do not understand our own, or if we are not prepared to engage in our own (Kaplan, 1999)? This is not to say that this is a revolutionary strategy. In fact, the activities that it entails would not transcend the conventional workshops, networks and e-forums. The main difference would be in the approach, which would focus on facilitating reflection on experience and learning from it, adding an assets-based element to capacity-building. Such an approach aims to establish the case for mainstreaming project learning and learning review.

With this in mind, it is interesting to look at an existing capacity-building initiative in participatory development communication to see how it could be adapted to suit the specific needs of the Arab region.

Case in point: Isang Bagsak

Isang Bagsak, a capacity-building and networking programme in participatory development communication, provides a good basis for the learning-to-learn approach. The programme combines the use of internet technology with face-to-face meetings to discuss themes related to PDC. The following points highlight key aspects of the Isang Bagsak programme and discuss ways of tweaking these aspects so that they incorporate more reflection and self-learning:

- Overall strategy: Isang Bagsak is considered a capacity-building and networking programme that largely relies on distance education in its modalities. A resource person introduces each theme by sending it in a message to all participating teams for their comment and synthesis. Although not intended, the discussions have taken the form of questions and answers, where the resource person stimulates discussion on each theme through trigger questions and the teams gather to prepare their response to the questions.
- To create a reflection/learning programme, this attitude must be instilled from the start. The focus of the programme should be presented as one that seeks to enhance knowledge on participatory development communication and to strengthen critical reflection skills. Participants should feel that this is an opportunity to think critically about their development work and to draw out lessons learned.
- Content and discussion: the Isang Bagsak programme relies on retrospective reflection on previous projects, or on aspects of current projects that have already occurred. While this is essential in order to draw lessons learned from experience, it does not necessarily enable participants to apply lessons learned to new situations.

    In order to facilitate this, participants would be asked to pursue two tracks when discussing each theme. One track would be to consciously reflect on the theme in retrospect through discussions of previous experiences. This would reinforce the perception that organizations have some experience of participation, albeit unconscious. The other track would focus on the discussion of each theme in light of projects currently being pursued. The idea here is to provide a small fund for each participating organization in order to select a natural resource management initiative jointly with communities. This would start at theme 3: involving the community in the identification of a natural resource management problem and its solutions. The fund would be nominal, only enough to kick-start an initiative with the community. From theme 3 onwards, participants would be encouraged to share discussion themes with their communities as part of a joint learning and reflection process and an informal evaluation mechanism.
- Facilitation: in the Isang Bagsak programme, a resource person is responsible for introducing the theme and commenting on each team's synthesis. This may create a sense that the resource person is the expert in control.

> While there should be a main resource person for each theme, participants should also take turns in co-facilitating themes. The resource person should only intervene in the discussion when necessary.

- Design: the framework of the Isang Bagsak programme is largely predesigned. Participants take control over the discussions through synthesized responses.

  In order to spur a learning and reflection environment, participants should have space to participate in the design of the programme. For instance, each participant can pose a critical question (for example, an ethical dilemma or an issue worth pondering) to each theme that can crosscut the discussions. Participants can also select for discussion the participatory dynamic that is of relevance and importance to them.

- Participants: several country teams have already participated in the Isang Bagsak programme. A core team in each country was responsible for convening the country teams and synthesizing their discussions. There was concern that the richness of discussions may have been lost in too much synthesis, especially when there were several teams in each country.

Participating teams should be able to transfer their knowledge to other organizations. This can be achieved through two mechanisms. Each organization participating in the programme should partner with another organization of less experience. In this way, one organization 'mentors' the other in implementing the joint initiative decided upon with the community. Each country would be represented through two teams – hence avoiding dilution of discussions. Another way to expand the programme beyond the participating organizations would be to generate materials such as simple guidebooks, training packages, frameworks and exercises that can be used by participants to disseminate knowledge to others. Compilation of such material would be the responsibility of a resource person other than the main facilitator. Such material would include case stories provided by the participants and their respective communities. Responsibility for using such material with other NGOs would be the responsibility of the participants. This notion draws on the experience of the Living University programme, initiated by Save the Children US in Egypt. The programme is based on peer-to-peer learning through NGOs. NGOs participate in an extensive training programme and either graduate to become a 'Living University' or a 'learner'. Living Universities then become ambassadors who transfer their knowledge to other NGOs, and the cycle continues in a cascading effect. According to the participating NGOs, learning from a peer NGO was a more positive experience than learning from a contracted consultant, as is the case with conventional training programmes.

Conclusions

To summarize, capacity-building in the Arab region should aim to bring out in a systematic way knowledge which may exist at a tacit, subconscious level within individuals or organizations. When we consider capacity-building in participatory development, we should also be considering capacity-building in learning. And consistent with the philosophy of participatory development of 'teaching a person how to catch the fish rather than providing the fish', capacity-building should also be about teaching how to learn, rather than providing ready-made learning.

Yet, it is not expected that Arab practitioners will jump at the idea of learning from experience. The issue of incentives and motivation needs to be carefully considered. Clear and tangible benefits must be demonstrated. Each practitioner will wonder about the personal gain from taking the time to participate in a learning initiative unless such time is factored into ongoing activities, perhaps as part of the 'dissemination' sub-item common to all development initiatives' budgets. Other incentives mentioned above include funding to start an initiative or the prospect of creating materials that are useful for training and other forms of knowledge dissemination.

Note

1   Qat is a sedative grown and consumed in Yemen and is considered an integral part of work and social life.

References

Al Sayyid (1993) 'A civil society in Egypt ?', *Middle East Journal*, vol 47, no 2, pp228–242

Freire, P. (1970) *Pedagogy of the Oppressed*, New York, Continuum

Ibrahim, S. E. (1996) *An Assessment of Grass Roots Participation in the Development of Egypt*, Cairo Papers in Social Science, vol 19, no 3, The American University of Cairo Press

Ibrahim, S. E. et al (1991) *Civil Society and Governance in Egypt*, Institute of Development Studies, University of Sussex, UK

Kanawati, M. (1997) *Consulting Egypt's Local Experts: Beyond Prince and Merchant*, The Institute of Cultural Affairs International, Pact Publications, New York

Kaplan, A. (1999) *The Development of Capacity*, United Nations Non-Governmental Liaison Service, Geneva

NEF (Near East Foundation) and CDS (Centre for Development Services) (1997) *Arabization for PRA Materials in the Documentation of the PRA Exchange Meeting: Challenging Practice Attitudes*, NEF and CDS, Amman

Saber, A. (2002) 'Towards an understanding of civil society in the Arab world', *Alliance Newsletter*

Siri, G. (2002) *The World Bank and Civil Society Development: Exploring Two Courses of Action for Capacity Building*, World Bank Institute, Washington, DC

Sulaiman, S. (1998) 'Democracy and Shura', in Kurzman, C. (ed) *Liberal Islam: A Reader*, Oxford University Press, New York

Implementing Isang Bagsak: Community-Based Coastal Resource Management in Central Viet Nam

Madeline Baguio Quiamco[1]

A young and fast-growing population and a robust tourism industry are providing the impetus for growth in the fisheries sector of central Viet Nam. As fish catch from the sea decreases in response to over-exploitation, aquaculture is responding to the demand for marine products. The culture of popular and valuable marine species such as lobster, shrimp, clams, crabs and preferred fish species in the lagoons and bays of central Viet Nam has become highly profitable, attracting investors and eventually creating the usual triumvirate

of problems that seem to follow demand-driven development: pressure on natural resources, displacement of local people from their livelihoods and inequity in the distribution of benefits.

Research aiming to understand the biological and social dimensions of fisheries resource management has been going on in central Viet Nam during the last decade. Between 1995 and 2001, as part of a research programme called the Lagoon Project, the Hue University of Agriculture and Forestry, the Hue University of Science and the Hue Department of Fisheries investigated the management of biological resources in the Tam Giang Lagoon. The research initiative also examined how global and national changes were affecting people's livelihoods in the province of Hue's coastal area.

The initiative, entitled Community-Based Coastal Resource Management (CBCRM) for Central Viet Nam, is a three-year project that seeks to utilize the results of the Lagoon Project and other research and development efforts to address coastal resource management problems in the region. It is being implemented by the three organizations involved in the Tam Giang Lagoon Project, as well as two others: the Research Institute for Aquaculture – Region 3 (RIA–3) and the Nha Trang University of Fisheries. Through the project, it is envisioned that approaches to coastal resource management throughout Viet Nam will be improved, that the complex livelihood connections will be better understood and that this new knowledge will influence local policy with a view to improving participatory management of coastal resources. Finally, a network of community-based resource management researchers will be initiated in Viet Nam to implement a capacity-building programme for all researchers.

Seeking grounds for partnership

The team's motivation to participate in Isang Bagsak[2] stems primarily from its desire to improve its scientists' capacity to use participatory development communication (PDC) to achieve coastal resource management goals. Team members saw in Isang Bagsak the answer to their need for 'social science' knowledge and skills. They expressed the belief that improved communication skills would enable them to conduct their people-related tasks more effectively and, ultimately, help the project to achieve its specific objectives, which are to:

- increase the capacity of the partner institutions for leadership in community-based coastal resource management in Viet Nam and to initiate a network of researchers as a means of providing capacity-building support for other Vietnamese institutions;
- understand the changes that have occurred in livelihood diversity, the coping strategies put in place by communities and local policy responses to those changes in lagoon resource use;
- identify and evaluate means and processes for scaling up fieldwork to include multiple communes using ecosystem-based management

principles while considering issues of larger socio-political organization and ecological aspects of living resources;

- assess and understand the impact of aquaculture on the livelihoods of traditional fishers; and
- explore options for improved participatory management of aquaculture and fisheries.

This rationale for participation convinced the Isang Bagsak South-East Asia coordinating group that its activities could, indeed, support the CBCRM's objectives. Isang Bagsak's other requisites for participation were also met: sustained connectivity, actual work going on in the community, and a willingness to apply PDC to their current work and to devote time to Isang Bagsak activities.

Initial results

Team members have been participating in cycle 1 of Isang Bangsak, together with three other teams in South-East Asia: the Forestry Administration in Cambodia, the Legal Assistance Centre for Indigenous Filipinos (PANLIPI) and the College of Development Communication (CDC) of the University of the Philippines at Los Baños, both based in the Philippines.

Collaborative implementation of introductory workshop

The introductory workshop for the Vietnamese team was held in Nha Trang, central Viet Nam on 20–22 February 2004. It brought together 16 fishery scientists from three implementing organizations: the Hue University of Agriculture and Forestry, the Nha Trang University of Fisheries and the Research Institute for Aquaculture – Region 3. During the workshop session aimed at levelling off expectations, it was understood that in participating in Isang Bagsak, participants expected to acquire the skills for working effectively with low literacy-level people, facilitating the activities of local people involved in development initiatives and helping them to solve their problems. They would also learn to plan PDC activities in community-based coastal resource management and acquire skills to involve more participants in community activities. Participants also expressed some concerns about some aspects of the programme and about PDC, which were discussed collectively.

E-forum

As far as connectivity and aptitude for the technical side of the e-forum are concerned, all three organizations have internet capability and the team members have access to it, although one team seemed to have less access than the others. At the e-forum briefing session during the introductory workshop, everyone demonstrated working knowledge of the internet. An hour after the briefing, the participants were finding their way around with the forum software.

Their problem was in responding to the forum questions, in constructing their messages and in posting them 'for all to see' since this needed to be done in English, the only language they had in common with the other teams from the Philippines and Cambodia. There was a general hesitation to post their thoughts because 'their English is not very good'. It was difficult for them to get beyond the salutation and greeting phase.

While some commented that the procedure was complex because it involved many steps, they also thought that the e-forum was a helpful and convenient way of communicating, and that it would be useful to them. To solve their language problem, they suggested a facility apart from the e-forum (and seen only by them) where they could construct their e-forum posting. They even suggested e-mailing their posting first to the country facilitator so that she or he could edit it before posting. A number of reasons explain this attitude:

- English is not spoken every day in Viet Nam, even by people such as scientists;
- posting on the e-forum can feel like one is publishing something (and therefore should be 'correct');
- offering one's thoughts and reacting to those of others in public is not easy for most Asians; these have to be done with care.

Learning participatory development communication through online sharing

Through their participation in the Isang Bagsak e-forum as the country team for Viet Nam, team members are increasing their understanding of the PDC concept and its application. For a team that began only recently with the major concern that it did not have sufficient knowledge of communication tools or a mastery of the English language, the Vietnamese team is doing well.

The team has maintained its participation in the e-forum, contributing richly to the knowledge exchange by posting its experiences in working with the coastal communities on the e-forum. At the same time, online exchanges of PDC experiences, ideas and analyses with the participating teams in the Philippines and Cambodia have broadened the Vietnamese team's understanding of the communication facet of its work with coastal communities. Isang Bagsak has introduced the perspective of PDC as a cycle of ten interrelated participatory steps. Using this new perspective as a framework, the Vietnamese team, along with the other participating country teams, is finding new meaning in the people-related work that it does in natural resource management. Through the e-forum, the Vietnamese team has shared the following experiences in PDC with the other teams.

Using communication to facilitate participation

The team has used a wide range of communication tools to facilitate participation, ranging from informal and formal conversations to training, and from participatory rural appraisal methodologies to mass media, specifically radio broadcast. The team explained that communication has helped to achieve natural resource management goals by motivating people to participate, clarifying cooperation among stakeholders, and convincing people and gaining their support, as well as through the enactment of laws and the issuance of policies, through the implementation of plans or projects, by mitigating boundary or resource-use conflicts, and by providing more choices or options. Communication cannot help, the team opined, when information is not updated to suit the changing context; when there is only one-way communication and no system for feedback, evaluation or monitoring; and when there is no transparency, causing people to lose confidence.

To be effective, researchers and development workers have to have certain skills and abilities. The team identified these as communication and presentation skills, the ability to motivate and work with people, and knowledge of local issues, concerns and socio-political conditions. Research and development workers, they said, must have planning skills, an open mind, fairness and foresight. They must also have cultural sensitivity, resourcefulness and an approachable personality.

Approaching the community

For the Vietnamese team, approaching a community involved first contacting the local authority or leadership, rather than going directly to the people. It required understanding the community's customs, beliefs and culture in advance. It also required understanding the local people and listening to them, including those whose ideas or opinions were in conflict with those of the team, those whose trust had been betrayed by other people from outside the community, and those with whom the team had limited contact. Approaching a community also meant clearly explaining the goals and objectives of the activity, sharing information with them, establishing relationships and becoming like local people themselves (i.e. eating, living with them and behaving as they did) to gain people's acceptance and pave the way towards working with them.

In doing this, the team encountered some initial difficulties. One such difficulty involved the community leaders, who would not allow the team to go into the village. The team solved this by spending time asking for help from higher-level officials – that is, from the commune or district authorities. Another difficulty the team encountered while initiating the project involved working time, which differed between the local people and the team. Team members worked with farmers and fisher folk. Fisher folk worked at night and rested during the daytime. Their solution was to adapt to the fisher folk's working schedule, highlighting the importance of flexibility and adaptation rather than bureaucratic efficiency among implementers.

In order to approach the community successfully and manage initial difficulties, the team needed more than what official documents could provide. The team felt that it required general information and secondary data, socio-demographic information on the people, and information on development projects and activities currently being implemented, as well as those planned for the next couple of years. The team gathered all this information by using participatory rural appraisal methods such as focus group discussions and timelining, and conducting household interviews.

The team described the people in the community as either fisher folk or farmers. Fisher folk are either fixed-gear fishers or mobile-gear fishers. In addition, the group of village leaders, women folk and the youth may also be identified. In general, the people in the community are very poor and have little education. They are friendly, but very conservative. The majority of them practise both fishing (fish capture) and aquaculture. Most farmers are rice farmers.

People in the community generally hold a very simple perspective about natural resources. They believe that the rice field is private, but that the fishing ground is a common resource. Thus, while they should not plant rice in someone else's rice field, they can fish and put fish cages anywhere in the lagoon. The team observed that traditional and modern media are used by the community. Activity organizers inform each household about important events, write news or conduct meetings; at the same time, people have access to mass media such as radio, television and the press (newspapers), as well as to posters and people's forums or exchanges of information. Since beginning to work in the community in 2003, the CBCRM team has shared with the people its knowledge and experience in community management and credit management; information about environmental and natural resource protection; and know-how in planning for aquaculture development.

Entering the community

The team approaches a community with an outline of an initiative's proposal. This document is the outcome of the team's discussions and states the overall objective of the initiative, the target area and the target group. However, specific objectives are still to be finalized based on the comments of the local people after the outline is presented to them. For example, one priority was for community activities that enhance the living standards of local people in environmentally friendly ways. Thus, research activities and interventions should follow this priority objective. But the team took time to build consensus in the community on what the priority activities are, and where/when these should be implemented.

People were asked to participate in identifying problems and their solutions. Through individual interviews and subsequent group discussions, they were asked what they had done to improve their livelihood, the problems they had encountered and how they thought these could be solved. In the group discussion, the team helped local people to share their experiences and ideas with each other, to hear one another's viewpoints, and to form a consensus on their problems, solutions and priority actions.

While the process had the potential to enable local people to understand the issues, form a consensus and decide on actions, in practice it still was constrained by several problems. Participation, it was observed, was influenced, first, by each participant's status and the culture of the group, and, second, by people's understanding of the project's activities and benefits. The team described the situation as follows:

> *A person's position in the community will either make him confident or not confident enough to participate in community activities. That is, a village leader is more confident than a poor person, even if he/she is normal. If a poor and not very educated person expresses his/her thoughts, it may not be treated as important as that of a well-educated villager's idea, even if the two ideas were similar. In a community we worked with, culture prevented women from participating. Women there were not used to talking at village meetings. According to an old concept: 'Women are inward, men are outward.' This prevents women from participating in such a meeting. Differentiation through social stratification (that is, small mobile versus large fixed fishing gear groups) also prevents the people from participating in the common action.*
>
> *Another problem that prevented participation was that our ideas to help people were sometimes not clear to them, so they did not realize the benefits from the actions to be undertaken. This limited people's participation.*

Testimonials about the initiative from people in the same circumstances and clear explanations of its goals, objectives and benefits helped mitigate these factors, which initially hindered people's participation.

Conclusions

Thus far, Isang Bagsak and PDC are helping CBCRM to enhance its effectiveness among its many stakeholder groups: fisher folk, fishery entrepreneurs and investors, policy-makers and implementers, other scientists and students of fishery. Through their interest in PDC, the team has demonstrated its capability to mobilize itself, collaborate and share resources. The team has maintained its active participation in the e-forum by sharing its experiences and ideas on the different participatory development communication themes discussed, although it still has to demonstrate its skill in reacting to the experiences of other teams and posting its insights.

Interaction with the other Isang Bagsak teams has also helped the Vietnamese team to identify its communication skills needs. Indeed, before the mid-term workshop was conducted in Viet Nam last August, team members were able to spell out the communication skills that they needed in order to improve their effectiveness in the field. On their request, the Isang Bagsak project team from CDC conducted skills development sessions on radio script writing and video production.

Through exposure to PDC concepts and its applications in the field in South-East Asia, the team is improving its understanding and knowledge of how it can utilize PDC for more effective research and development work in natural resource management. Isang Bagsak implementers are helping the team to nurture this understanding and knowledge by providing an atmosphere for participatory learning – that is, by demonstrating resolve, clarity of purpose, respect for differences, transparency and a genuine intent to empower through capacity-building. Consequently, this should build a sense of responsibility for Isang Bagsak activities, provide mutual trust and enhance the capacity for participatory development communication all around.

Note

1   This chapter was written with inputs from the community-based coastal resource management team at the Hue University of Agriculture and Forestry in Viet Nam.
2   Isang Bagsak is a learning and networking programme that aims to improve communication and participation among natural resource management researchers, practitioners, communities and other stakeholders, and to provide communication support to development initiatives aimed at helping communities overcome poverty.

Building Communication Capacity for Natural Resource Management in Cambodia

Jakob S. Thompson and Mario Acunzo

The government of Cambodia has established as a priority achieving and maintaining food security for its rapidly growing population. Existing national policies promote the sustainable multiple-use management of natural resources at the community level. However, this cannot be accomplished without information and communication interventions aimed at altering the negative attitudes and reducing the unsustainable and damaging practices of natural resource users. This, in turn, requires a strengthened national institutional capability to design and implement targeted information and

communication interventions in support of local community natural resource management plans and efforts. Failure to establish this capacity will result in diminished institutional capacity to prevent and mitigate environmental degradation, with resulting negative impacts on agricultural productivity and national food security.

Objective and activities

The overall objective of the United Nations Food and Agriculture Organization (FAO) initiative entitled Information and Communication for Sustainable Natural Resource Management in Agriculture is to contribute to the improvement of natural resource management in Cambodia through building national capacity in the systematic design and use of information and communication strategies, methods and materials. The primary activity undertaken by the initiative is the training of central- and provincial-level staff from the Ministry of Agriculture, Fisheries and Forestry and the Ministry of Environment in the theory, design and use of participatory development communication (PDC). During the 36-day training programme, the 19 participants who have been trained so far carried out the design and preliminary practical implementation of a strategy for PDC with villagers in two pilot sites. This learning-by-doing process included participatory analysis, training of villagers, and material design and production, as well as monitoring and evaluation for the improvement of agricultural and fishing practices. The training programme was carried out by a team of trainers from the College of Development Communication (CDC) of the University of the Philippines at Los Baños.

Training strategy

The project's capacity-building strategy is based on building field experience through in-service training and learning-by-doing strategic planning, implementation, monitoring and evaluation in local-level pilot sites. The step-by-step training approach used in this initiative can be summarized as follows:

- establishment of a project coordination unit and a communication team, comprising 16 communication unit staff from the two ministries;
- training of eight central-level communication team members in the design, implementation and evaluation of information and communication strategies;
- selection of pilot sites;
- study tour for two selected team members to the University of the Philippines at Los Baños;
- theoretical and practical training of team members through participatory situation analysis; planning, monitoring and evaluation; material development; and implementation of the plan, in collaboration with 73 natural

resource users, with a focus on identified natural resource management issues;
- training of team members in equipment handling, operation and care;
- training of team members in cost recovery.

So far, the communication team has undergone five training workshops for a total of 36 days, and another three workshops are planned to be held before the end of the project.

Achievements

The impressions and feedback gathered at all levels strongly suggest that the approach has led to a notable improvement of knowledge, skills and behaviour of both government staff and villagers. This is most evident in the way in which the work is being carried out by the communication team and the way in which people have changed their lives in the pilot sites. However, the institutional impact of the project is equally important.

Project impact

Team members have expressed the fact that the project has helped to improve their technical skills and the way in which they work with people in the communities. They say that they now handle desktop publishing and video editing effectively. At the same time, their appreciation of peoples' involvement in the development and implementation of information and communication initiatives has improved.

In the communities, people say that the activities have led to improved practices and increased awareness, such as improved production in pig raising and less cutting of flooded forests. People also raise new issues, such as chicken breeding and precautions relating to avian flu, as communication challenges for the community. This illustrates their appreciation of PDC as a way of dealing with the problems that they face.

A project coordination unit has been operational since the onset, based on the will and ability of two ministries at the provincial and central levels to join forces. For the future, this cooperation has demonstrated the benefits of working together to tackle PDC challenges as the strong links between staff in the two ministries now form a base for continued cooperation. The project has also made a valuable contribution in the form of the Cambodian government's acknowledgement of the importance and potential benefits of planned and targeted information and communication interventions when dealing with complex natural resource issues. To go with it, this initiative can provide examples in the form of training and field-level methodologies, as well as communication tools and materials. These aspects are also likely to have improved the conditions under which community-based natural resource management will develop into an integral part of the Cambodian government's work to increase food security in the future.

Equipment

The project has equipped the ministries with a digital video production and editing system, a digital still camera and a desktop publishing system that is in constant use, and trained staff members are becoming more and more confident and effective in the use of high-tech tools for media production. At the provincial level, the communication team has also received the same equipment. However, the opportunities to practise are not as frequent and the equipment is not as much in use as is the case at the central level, where the equipment is kept and used in an easily accessible area of the office, also under constant supervision and maintenance.

Additional capacity

The training capacities attained by the communication team were demonstrated and further strengthened when the project provided strategic communication training to a FAO community-based natural resource management initiative in Siem Reap Province. The communication team trained staff in designing and producing multimedia tools as part of a one-week training course. The outcome of the training and feedback from participants supports the impression that this is a task that the communication team can fulfil in times to come. Currently, the communication team is also providing training and technical support to the Special Programme for Food Security and to the Community Fishery Development Office in order to implement a communication strategy design and implementation process. As a result, the communication team is able to strengthen its skills and further develop its capacity as trainers, which is likely to constitute a major part of the support that it is intended to provide to programmes or projects in the future.

Current situation

Today, an important step has been made for PDC to contribute to sustainable natural resource management in Cambodia. As mentioned earlier, the communication team is currently providing support to two natural resource management programmes to build and implement information and communication components. Besides providing team members with an opportunity to practise and perfect their newly acquired skills, these large and highly visible natural resource management programmes are able to display the quality of materials and products of the communication team as a service provider. This, again, can form the basis for the team to become a responsive and capable service provider in mainstreaming and scaling up PDC in the natural resource management sector and in the evolving development scene in Cambodia.

Constraints faced by the initiative as a result of ...

... relying on foreign training capacities

So far, the main constraints faced by the project have been that project activities have more or less come to a halt when the international consultants were not in the country to provide training or to push for activities. This has been dealt with by hiring designate national staff to ensure the continuity of activities between missions. A general lack of operational budgets, for example, to travel and pay for materials is making it difficult for the communication team to work and practise the new skills. Relying on out-of-country training capacities is also making continued backstopping and follow-up difficult – which is something the team members clearly express a need for, especially in relation to the use and maintenance of computer-based tools.

... language

With regard to the actual training, Cambodia, as many other countries, poses a challenge when it comes to language. The use of foreign training capacities hampers the effectiveness of the training, and there is a constant risk that important points are lost in translation.

... skills and attitude of trainees

The prior attitude of fieldworkers is also something that one needs to take into consideration when designing and implementing training schemes. The trainees often have a more direct approach to teaching, which often contradicts participatory approaches to learning. The relatively low level of fieldworker's natural resource management knowledge also influences the will to assume an important underlying principle of participation – namely, flexibility. An underlying reason for this is probably that handing the leading role to people sometimes amounts to stepping into unpredicted fields of knowledge that are alien to the field worker. This again poses a threat to their authority, which is probably why trainees tended to focus more on the underlying subject than on the communication process, at least during the early stages of the training programme.

Lessons learned

Selection and levelling of content

The complexity and wide-ranging content of the project's training programme has proven to be a challenge. The trained team was comprised of staff from different levels and different backgrounds. Some had prior video production training and computer skills, some had experience working with communities, but few came to the training with a mix of the two. These two groups expressed different training needs some time after the training: 'practise-oriented' staff wanted to further strengthen their community organization skills and general ecological knowledge, whereas technology-oriented staff clearly expressed the need to learn more about the use of new software or about the technological part of development communication. Training needs appeared to be related to what team members already knew. There was a tendency to want to strengthen existing knowledge and to specialize, rather than to adopt the wide range of skills that a participatory development approach requires of the facilitator.

There is a need to adapt the curriculum to local conditions and to adjust the amount of information to be provided to intended learners accordingly. It is also necessary to assess the usefulness of high-tech solutions. As an example, team members now wish to learn how to use different video editing software since a local TV station is relying on them to provide material according to a certain format. A thorough assessment of the prior knowledge of the trainees and the framework for their work must be used to shape a training programme.

Allowing for practice

The high-tech equipment provided by the project is expensive, and people tend to treat it with care – or don't use it at all. If people are to become effective users of tools such as digital cameras or desktop publishing devices, they have to practise these techniques on a regular basis. This is particularly true in the provincial office, where the communication team is waiting for additional training before they put the tools to use. It is therefore important to emphasize that one should not sit and wait for opportunities or actual tasks to put these important tools to use, but to continually practise. In this way, skills are strengthened and can be ready when the need arises. It is also better to have a worn-out camera that many people can use, than to have an unused camera that no one knows how to use effectively.

Sustainability

In order for new knowledge and skills to become part of local capacity, there must be practical opportunities, as well as personal and cooperative adaptation. In this case, most targeted trainees were from central-level government units, whose aim was to create central-level capacity and to ensure the institutionalization and visibility of project activities. However, the staff had few opportunities to gain field-level experience unless with external support since operational budgets for travel are commonly lacking in government institutions. Consequently, central-level staff members engaged in TV programme production and computer-based work, while the necessary conditions for the field aspects of PDC to develop into practical skills were lacking. At a provincial level, there is never a lack of opportunities to work with communities and to practise the field aspects of the training programme; but the trainees have expressed the need for greater post-training support at the implementation stage, a lack of which is preventing them from further using the high-tech skills developed by the project, such as digital media production. One solution is to train staff at different levels on complementary skills and to have them work on a participatory, technical or cooperative basis depending upon their level of government. A second option is to create opportunities for field staff to practise technical skills and *vice versa* in order to lay the basis for cost-recovery schemes, thus facilitating independent field travel or backstopping in the future.

Implementing Isang Bagsak:
A Window to the World for the
Custodians of the Philippine Forests

Theresa H. Velasco, Luningning A. Matulac and
Vicenta P. de Guzman

Isang Bagsak South-East Asia is an enabling programme in more ways than
one. The implementation of this networking programme, through its basic
strategy of participatory development communication (PDC), is giving rise
to experiences and insights on involving the community in development
work through communication. More significantly, this involvement hinges
on the use of information and communication technology – a 21st-century

tool whose applications for development are currently gaining ground in developing countries.

The programme is being implemented in the Philippines by the College of Development Communication (CDC) of the University of the Philippines at Los Baños and its partner in the first cycle, Tanggapang Panligal Para sa Katutubong Pilipino, or the Legal Assistance Centre for Indigenous Filipinos (PANLIPI). The decision regarding the Philippine partner for the Isang Bagsak learning network was not an easy one to make. There were many mainstream organizations with vast experience in natural resource management. Resource constraints also figured prominently on the choice of partner.

In the end, the CDC posed one question that made all the difference: who would benefit most from the Isang Bagsak experience? The following rhetorical question also tipped the balance in favour of PANLIPI: why not share the learning from Isang Bagsak directly with the indigenous peoples, the custodians of Philippine forests?

According to the Indigenous Peoples' Rights Act of the Philippines, indigenous peoples refer to a group of people or homogeneous societies identified by self-ascription or ascription by others who have continuously lived as organized communities in a communally bounded and defined territory and who have, under claims of ownership, since time immemorial, occupied, possessed and utilized such territories sharing common bonds of language, customs, tradition and other distinctive cultural traits. Indigenous peoples have, through political, social and cultural inroads of non-indigenous religions and cultures, become culturally differentiated from the majority of Filipinos. Indigenous peoples will include people who are regarded as indigenous on account of their descent from populations who occupied the country before colonization and who retain some or all of their own socio-cultural or political institutions.

A showcase of synergy

PANLIPI is an organization of lawyers and advocates of indigenous peoples' concerns. Established in 1985, PANLIPI primarily aims to assist the indigenous peoples in their struggle for the recognition of their rights to their ancestral domains, culture and traditions, and other basic rights. The end goal is to empower the indigenous peoples to the fullest so that they can actively participate in every aspect of Philippine society. PANLIPI's mandate includes the provision of development legal assistance; legal education and outreach; institutional capacity-building; ancestral domains delineation; and resource management planning.

The networking among Isang Bagsak, CDC and PANLIPI may be viewed as a synergistic one right from the start. From PANLIPI's viewpoint, the promotion of the rights of indigenous peoples to their ancestral domains and the issue of natural resource management go hand in hand. The indigenous peoples are, after all, the best caretakers or stewards of the natural resources mainly because their socio-economic system is sustainable and not destructive.

For its part, CDC found in PANLIPI a partner who could best make use of the very tenets of development communication – that is, the use of communication towards improved socio-economic growth of a community that makes for social equity and the larger unfolding of individual potential. Through Isang Bagsak, the indigenous peoples of the Philippines are given their window to the world of learning and are sharing their own knowledge, insights and experiences with members of the network from various parts of the globe. Likewise, the other members of the network are allowed access to the rich heritage of peoples hitherto unwired to the Isang Bagsak network.

Lessons learned

Working with PANLIPI is, indeed, a work in progress that is bringing forth a lot of valuable insights for researchers and development workers in general. The introductory stage of the partnership was already fraught with lessons, foremost of which were the ethical considerations. For one, is it ethical to bring in the indigenous peoples to a programme that is quite alien to them in terms of the tools (information technology) to be used? Second, will they agree to become part of the undertaking, and how do we engage their participation?

Mutual respect and trust

There was conscious effort on the part of the CDC to show respect and to gain the indigenous peoples' trust at the start of the project. Before any agreement was inked, the project team, with PANLIPI's help, made a point of observing two things:

- understanding the cultural context; and
- securing the indigenous peoples' free and prior informed consent.

Henceforth, a series of consultations with the indigenous peoples who would be involved in the learning network was undertaken. The CDC team, together with the PANLIPI lawyers, literally sailed, crossed rivers and hiked through mountains to dialogue with the elders of the cultural communities. This was part of securing their free and prior informed consent, in keeping with the PANLIPI protocol and the nature of participatory development communication. These are two critically important values that many development workers seem to be taking for granted, but which the Isang Bagsak experience is bringing to the fore in this undertaking.

Genuine dialogue between and among the CDC, PANLIPI and the indigenous peoples paved the way for the participation of people from four areas: Mangyans from Mindoro; Tagbanuas from Palawan; Kankanaeys from the Cordilleras; and Aetas from Zambales. Iterative consultations leading to consensus-building were held. First, the CDC team conferred with the PANLIPI staff and lawyers assigned to the areas. Once PANLIPI's cooperation was ensured, dialogues were held with indigenous elders and leaders, and eventually with the members of the community.

The series of dialogues highlighted the following points:

- Knowledge sharing: Isang Bagsak offered an opportunity for the PANLIPI network and the indigenous peoples in the four areas to not only learn from others and from one another about natural resource management and PDC, but to contribute their own knowledge to the network.
- Assurance that the indigenous knowledge systems (IKS) would be respected: the indigenous peoples voiced a strong concern about the possible piracy of their traditional knowledge, as was their experience in the past. They were wary of outsiders 'stealing' their indigenous knowledge in the guise of development work. Prior experience of being exploited by researchers from outside (foreign and local) was a big stumbling block to securing the indigenous peoples' free and prior informed consent. Team members took pains to explain that nothing that did not come from the people, through their representatives in the programme, would be posted on the website. This was, in essence, a manifestation of respect for intellectual property rights.

Relevance to indigenous peoples' concerns

The indigenous peoples' decision to participate in Isang Bagsak rested largely on the realization that the skills they would develop would be useful to the concerns that they were advocating, such as the Ancestral Domains Sustainable Development and Protection Plan and other important provisions of the newly passed Indigenous Peoples' Rights Act.

PANLIPI participants saw the Isang Bagsak experience as an excellent opportunity to interact with professionals from other countries. Through the sharing of experiences, one important insight surfaced: the affirmation that all along, PANLIPI and the indigenous peoples have been practising the principles of development communication, in general, and those of PDC, in particular. The training that the PANLIPI participants have undergone in the course of their participation in Isang Bagsak has also afforded them the opportunity to write and communicate their ideas clearly.

Information technology for indigenous peoples

Technology could be an awesome development for indigenous peoples, only very few of whom have been initially exposed to computers, much less to the internet. Part of their initial reluctance to be part of Isang Bagsak was apprehension about the technology itself. Connectivity/accessibility also loomed as a potential problem.

To address these concerns, PANLIPI, for its part, has begun sponsoring a series of training sessions on information technology for the indigenous leaders involved in Isang Bagsak. CDC's information technology team trained

Iraya Mangyan leaders in Mindoro during February 2004. A user-friendly sourcebook on using the computer and the internet was developed by the CDC especially for this training. The sourcebook, in turn, would be useful in future information technology training for indigenous peoples. In addition, PANLIPI has been very supportive in enabling the indigenous peoples to have access to computers and the internet.

CDC and PANLIPI team members agree that one of Isang Bagsak's 'by-products' is the gradual mainstreaming of the indigenous peoples into the digital age.

PDC through the net: Teamwork at its best

By the time the project was in full swing, the PANLIPI–Isang Bagsak network had 45 members, 9 each from the 5 indigenous groups. A network within a network was created by the partners. As the saying goes: *together, everyone achieves more.*

The members were selected by the indigenous communities themselves on the basis of their knowledge and skills on natural resource management issues. Other criteria were also included in the list of qualifications. Prospective team members had to be knowledgeable about natural resource management and indigenous knowledge systems and practices. They had to be capable of facilitating community participation. Finally, they were also expected to understand development work and to be knowledgeable and skilled in promoting a rights-based approach to development.

The resulting team was dubbed *Limang Tagupak*, a Mangyan term representing victory. The name also signifies a five-point star, representing the configuration of the five-team network at the PANLIPI level. The star is also symbolic of a guide.

The sharing and learning process on PDC in natural resource management was carried out in the following manner:

- Local teams undergo orientation and training on Isang Bagsak.
- Local teams participate in the forum:
  - The CDC posts theme questions.
  - PANLIPI National Capital Region translates theme questions and sends them to local teams.
  - Local teams discuss the theme questions, formulate answers and send them to PANLIPI National Capital Region.
  - PANLIPI National Capital Region translates answers and posts reply.
  - PANLIPI National Capital Region downloads and translates country comments and sends to local teams.
- Local teams discuss country comments and make their own reflections, which they send to PANLIPI National Capital Region.
- PANLIPI National Capital Region translates and posts reflections.
- The CDC resource person synthesizes.
- PANLIPI National Capital Region translates the synthesis and sends it to the local teams.
- Local teams discuss the synthesis and extract the learning.

The process appears to be a tedious one. True enough, a number of problems hampered the implementation of Isang Bagsak at the PANLIPI level. For one, translations into English took up quite some time; hence, delays in posting were inevitable. Delays were also due to the team members' tight work schedules or their preoccupation with economic activities, leaving them less time to participate in the forum. Access to the internet (lack of equipment, telephone lines, internet terminals/connection) constituted the hardware part of the limitations. Financial costs of consultations and regular meetings, as well as internet rental, posed very real problems as well. So did forces of nature, such as bad weather preventing team members from meeting.

PANLIPI's response to the resource-related problems mirrored its commitment to the partnership. PANLIPI put up counterpart funds for training indigenous peoples on basic computer use and internet navigation, as well as hardware for internet accessibility. It forged cooperative undertakings with non-governmental organizations (NGOs) and other friends to boost its investment in equipment and internet access. To facilitate meetings among the team members, PANLIPI made arrangements for Isang Bagsak activities to coincide with regular meetings of indigenous peoples' organizations.

The foregoing are just four of the insights arising from several months' implementation of Isang Bagsak in the Philippines. Many more could be drawn from the researchers' and the indigenous peoples' experiences as they go through the cycle in the remaining months. Isang Bagsak South-East Asia is a work in progress.

VI

Conclusion

Facilitating Participatory Group Processes: Reflections on the Participatory Development Communication Experiments

Chin Saik Yoon

The case studies in this book are a special collection of attempts aimed at putting people first in the development process. They are challenging but very worthwhile efforts intended to find ways of using development communication

tools and methods in decidedly participatory ways that advance the collective priorities of communities as determined by themselves.

These experiments are grounded in issues related to natural resource management. This is a sector for which participatory approaches seem most apt. Natural resource management is unlikely to be effective without the involvement and support of the communities. Natural resources such as land, water, air and forests are vested in the 'commons' – spaces and locations that are open to all people and often protected by nobody – belonging to everyone, yet cared for by none.

Successfully facilitated participatory approaches enable communities to assume collective ownership of the commons and to manage them in a way that safeguards the people's long-term interests. In the absence of such participation, the depletion of natural resources is hastened as people rush to extract maximum use and benefit of these resources before others do the same. Participatory approaches may help to replace destructive competition with sensible cooperation to preserve rather than plunder resources.

The triggering of such cooperation, in turn, rests on our abilities to strengthen and activate our 'internal commons'. These invisible commons comprise our cultures, values and sense of community. They are the elements that ultimately inspire people to set aside desires for selfish short-term benefits in return for the long-term security of the community. Degradation of our internal commons will lead to the destruction of the physical commons and *vice versa*.

Focusing on facilitation

Natural resource management is a challenge to most existing development-promoting strategies because they have largely been designed to advance short-term interests. Such promotional and marketing techniques understandably fail when the ultimate objective is for people to consume less and to set aside their self-interests and preferences for the greater good of the community, which in turn ensures the sustainability of future generations.

The participatory method attempted in most case studies reviewed in this book is participatory development communication (PDC). The opening chapter of this book by Guy Bessette, 'Facilitating Dialogue, Learning and Participation in Natural Resource Management' details the multifaceted work of communicators supporting community participation. It also provides a wide-ranging review of the tools, methods and strategies that the case studies in this book applied. The overarching role of the PDC practitioner is of a facilitator of participatory group processes that serve communities in building awareness and reaching consensus. This concluding chapter sets out to discuss and identify the group dynamics that the PDC practitioner facilitates.

The focus on facilitating participatory group processes rather than implementing one-way communication strategies is what sets PDC apart from classical communication for development (ComDev). In the latter, practitioners apply an integrated set of strategies and tools to mobilize people to meet a set of predefined development goals.

In PDC, facilitators align themselves with a community and help to organize activities and facilitate processes that bring people together to strengthen their sense of community, sharpen their awareness of shared aspirations and problems, and undertake collective action to realize what they aspire to be. The collective approach builds confidence in people, while at the same time making their action more effective through pooling resources and sharing of risks.

Because social communication is one of the principal elements that bind communities, PDC practitioners are able to enhance the usefulness of their facilitating role with their expertise in communication. In helping communities to build their capacity to communicate, they not only strengthen the ability of communities to organize themselves, but also to reach out to others within and outside their communities.

Facilitating key processes

Some of the key communication and group processes that PDC practitioners focus on facilitating are summarized in Box 6.1.

The processes that PDC practitioners concentrate on facilitating may be clustered into four categories:

1 communicating effectively;
2 creating knowledge;
3 building communities; and
4 enabling action.

Conventional ComDev efforts usually concentrate on the first and last clusters (communicating effectively and enabling action). PDC covers two additional clusters (creating knowledge and building communities) that aim to self-empower people through augmenting and validating their knowledge of critical issues and subjects that affect their lives, and through forging strong alliances among people, groups and communities so that they can consult and act effectively together in order to address problems and realize aspirations.

It is the facilitation of these two additional sets of group processes that provides PDC with its participatory bias and its potential to support initiatives that are sustainable in the long term. Many development efforts have been undertaken in the past by focusing on effective communication and enabling of action; however, this narrow focus often leaves such efforts vulnerable to eventual failure. They fail because people lack ownership and relevant in-depth knowledge to assume control of activities in the long term and, more importantly, because they lack the sense of a community.

Box 6.1 *Key participatory development communication processes*

Key participatory development communication (PDC) processes include the following:

- Effective communication:
  - self-expression;
  - listening;
  - understanding.
- Creating knowledge:
  - sourcing information;
  - tapping indigenous knowledge;
  - processing and validating information;
  - sharing knowledge.
- Building communities:
  - building trust;
  - managing conflict and competition;
  - forging partnerships;
  - reinforcing self-identity;
  - reflecting on the past and present;
  - visioning the future;
  - affirming values;
  - adjusting values;
  - enabling transparency in decision-making;
  - sharing benefits.
- Enabling action:
  - identifying problems;
  - evolving solutions;
  - nurturing a sense of guardianship of the commons;
  - managing expectations;
  - taking stock and pooling resources;
  - sourcing complementary resources;
  - advocating to stakeholders;
  - mobilizing for action;
  - evaluating action;
  - iterating and refining action.

Creating knowledge

Although ComDev efforts often devote quality attention to the dissemination of information, knowledge is sometimes not created among the people who are targeted with such information. Dissemination efforts also rarely take into account the rich knowledge base resident within communities. This is especially significant in natural resource management where indigenous knowledge is just as important as scientific information. Indigenous knowledge is distilled over generations through people's close observation of their environment and is therefore particularly valid for the long time spans involved in managing natural resources.

N'Golo Diarra from Mali retells his unforgettable encounter with the 'governess of the seasons' of a distant village in his chapter 'The Old Woman and the Martins: Participatory Communication and Local Knowledge in Mali' in Part III. This woman had inherited an important piece of indigenous knowledge from her father, the great traditional healer of the village. He had observed that the purple martins build their nests on tree branches that were always above the floodwaters of the river that ran by the village. He shared this critical piece of knowledge with his daughter just before he died. Armed with this precious element of indigenous knowledge, she became the much-revered governess of the seasons, helping farmers in her village to decide where and when to plant their crops, as well as where to pasture their livestock during the coming season.

PDC attempts to facilitate the fusing of indigenous knowledge with scientific information. So, instead of starting with scientific information, PDC approaches set out to empower people by validating indigenous knowledge, which they know intimately, before progressing to the introduction of scientific information. Participatory approaches adopted in filtering, processing and applying new and existing information help to create knowledge that allows people to take charge and make decisions.

The importance of processes that create knowledge is best appreciated in the chaos–wisdom continuum:

$$\text{Chaos} \rightarrow \text{Data} \rightarrow \text{Information} \rightarrow \text{Knowledge} \rightarrow \text{Wisdom}$$

The continuum begins with *chaos*, fragmented and disorganized data that is not of use to people. *Data* may be clusters of numbers and visual observations that have been processed and made ready for use. *Information* refers to data that has been organized into meaningful chunks that provide meaning to people. Information becomes *knowledge* when it has been successfully communicated to and understood by the people. 'Knowledge is the product of information plus thought and ideas. It implies a value judgement because knowledge marks the processing by a human of useful and relevant information' (Green et al, 2005). *Wisdom* occurs when knowledge is used to make sound judgements.

The strategies often adopted by conventional communication approaches focus just on the dissemination of data and information to people. Frequently absent are the participatory group processes that permit people to convert these raw inputs into useful knowledge that they can use as a community.

Building communities

The sense of community varies across communities. It may be very strong among rural communities which are bound by cultural, religious and social ties forged over numerous generations. Or it may be weak, as among landless communities comprising a high percentage of transient members.

PDC approaches try to tap into existing community processes whenever possible in order to strengthen rather than undermine communities. Facilitators will therefore adopt traditional and folk media, and communicate via existing channels and networks to which they have access.

In the case of groups with weak ties, quality efforts need to be devoted to bringing people together to forge a sense of community. This is often attempted in conjunction with efforts to create knowledge. Here, information and data are shared, and people are helped to process the information and data in groups so that they not only create knowledge for themselves but discover, at the same time, the many similar experiences and problems that they share, building a sense of community in the process (see the following section on 'Evolving participation in West Java' for a case study of this approach).

In other cases, stress and conflict created by competition for depleted supplies of natural resources can seriously test social and community ties. Karidia Sanon and Souleymane Ouattara from Burkina Faso provided us with such a case from the Nakambé Basin of Burkina (see 'Water: A Source of Conflict, a Source of Cohesion in Burkina Faso' in Part III). The areas surrounding the wells in Silmiougou, a village in the heart of Burkina, had degenerated into 'boxing rings' where women would compete each day for limited access to water pumps and water. The competition had become so fierce that fights broke out among the women every day, when they would smash each other's water jars. The authors reflect on how the PDC team was able to help the community to find resolution to the conflict via carefully facilitated chat sessions, which were primed with PDC tools, such as forum theatres, participatory video and radio, and posters.

Tools and methods

ComDev's previous emphasis on effective communication and enabling action has led to more tools and methods being developed in these two areas. This is apparent in the cases presented in this book where we see well-tested communication strategies being competently applied.

The potential of PDC becomes apparent when practitioners go beyond these strategies and begin to innovate tools and methods that support and facilitate people in the other two areas: creating knowledge and building communities. PDC practitioners have considered various group processes in their search for appropriate tools. Participatory media methods, participatory rural appraisal (PRA) (Chambers, 1994) and Visualization in Participatory Programmes (UNICEF Bangladesh, 1993) have all contributed effective techniques and methods for PDC.

The experimentation with novel tools for facilitating participatory communication processes undertaken within the cases reviewed in this book is significant not only for the efforts of the PDC practitioners, but more importantly for the active participation of the thousands of people from the villages and communities where these cases took place. The 'new' tools, techniques and methods reported in this book are therefore as much the innovations of the villagers and development fieldworkers as they are of the PDC practitioners who wrote up the cases.

Evolving participation in West Java

The case reported by Amri Jahi in Indonesia (see 'From Resource-Poor Users to Natural Resource Managers: A Case from West Java', in Part III) is important partly because it is the longest running development project studied in this book. Stretching over more than 15 years, the initiative was sustained during much of this time by the communities and the PDC team on their own. More significantly, it began in Indonesia during an authoritative regime when participation was not encouraged and the development model was dominated by top-down, expert-to-farmer initiatives. Although the case reads very much like a classical ComDev project, starting with a traditional agricultural extension intervention, it evolved gradually and subtly over a decade and a half into a vibrant PDC initiative managed entirely by the landless goat herders who succeeded in engaging constructively with stakeholders in various government agencies. The natural resource management problem of eroding riverbanks that had previously set landless communities in conflict with the authorities has been amicably resolved to the interest of everyone. The future of the river valley is secure and so is the livelihood of the goat herding communities.

The switch over to PDC began with the involvement of farmer leaders and farmer cooperators in monthly meetings to share information on how to care for the sheep. These meetings aimed not just to disseminate information, but also to build a strong knowledge base within the community. The meetings also served a second but more important goal of building a sense and purpose of community. An immediate task before the community was the distribution of lambs that were returned to the community in fulfilment of earlier agreements by farmers to repay their loans of breeding stock in the form of the offspring of their animals.

The project team discovered one of the most effective PDC methods early on in its work when it brought together farmer leaders to share their experiences with farmers from other villages. The farmer-to-farmer communication proved to be more effective than earlier methods attempted by the researchers and extension workers. The participatory approach was proved to be sound.

Challenges continue to face the community 15 years on. The PDC facilitators have noticed that the quality of the lambs returned to the community has declined recently. This threatens the sustainability of the revolving stock of young animals that enable other landless villagers to take part in the scheme.

Participatory videos and media in Lebanon and Uganda

Participatory video is one of the earliest PDC tools. It was developed on the Fogo Islands of Canada around 1967 by Don Snowden, then director of the Extension Department at Memorial University in Newfoundland. This method came to be known as the Fogo Process (Williamson, 1991) and has since been adapted and used around the world. In the case contributed by Shadi Hamadeh and his colleagues, we see the Lebanese application of this method to manage conflict and to provide women with a medium for empowerment (see 'Goats, Cherry Trees and Videotapes: Participatory Development Communication for Natural Resource Management in Semi-Arid Lebanon', in Part III).

The project in Lebanon relies mainly on interpersonal methods in its work. Additional PDC tools were used when interpersonal approaches turned out to be unsuitable. This was the case when people who were in conflict refused face-to-face meetings as a means of resolving their differences. The facilitators decided to interview the people involved in the conflict separately in order to obtain their perspectives on the issues. A video comprising the individual interviews was then screened to a meeting of all parties involved in the conflict. The video succeeded in restarting stalled face-to-face discussions on solutions to the people's differences. The meeting itself was videotaped. The new footage of the meeting, together with the individual interviews, was made into a new tape and was screened at another meeting involving others from the community in order to expand discussions on resolving the conflict.

The Lebanese facilitators then produced a video that featured interviews with women highlighting their economic productivity in a pastoral society. The video was shown to both men and women. The women who were featured in the video felt empowered after the viewing. This is the 'mirroring' effect discovered in the Fogo Process. The video had served as a mirror of their community demonstrating and recognizing the women's role within it.

Although the Fogo Process started as a 16mm documentary production before evolving into video, the process can also be applied to a whole range of modern media, ranging from radio to the internet. This is a promising area for research and development among PDC practitioners in the years ahead.

The case from Uganda contributed by Nora Naiboka Odoi features multiple adaptations of the Fogo Process, but using still cameras, brochures and posters in addition to video (see 'Growing Bananas in Uganda: Reaping the Fruit of Participatory Development Communication', in Part III and 'Communication Tools in the Hands of Ugandan Farmers', in Part IV). The iterative processes followed in developing the video and print material may look confused at first reading, but they served the very important purpose of processing information to create knowledge about banana cultivation, in addition to PDC materials production.

Community radio in Ghana

Kofi Larweh's case about Radio Ada, Ghana's first community radio station, describes how the medium taps into the oral traditions of the community served by the station (see 'And Our "Perk" Was a Crocodile: Radio Ada and Participatory Natural Resource Management in Obane, Ghana', in Part IV).

In community broadcasting, the listeners determine the content that is put on air and take turns in producing the broadcasts. This is an adaptation of the Fogo Process to the medium. Radio Ada staff see themselves as facilitators: they record many of the programmes in the villages, rather than in the studios. The villagers decide which topics they want to put on tape and the way in which they are presented to the listeners.

This case discusses the role played by the station in people's efforts to dredge a clogged 10km long river. It was a daunting task. Radio Ada first provided people with a medium to mobilize others living along the river to join in the huge efforts of dredging the waterway. The station then became a source of encouragement as the back-breaking work progressed. The names of the villagers who took part in the dredging were announced over the air, thereby recognizing their contributions, while at the same time encouraging others to pitch in.

The community succeeded after four years of hard work and triumphantly reopened the waterway to boats and fishing after nearly 40 years of neglect. Radio Ada had stuck with the people through these years. The unclogged river could now also channel water to the irrigation canals of riverside farms, and crops could once again thrive. The people had by now discovered the power of radio and began using it as an advocacy tool, raising issues about the environment and their livelihood over the air. They had not only rediscovered a waterway, but a communication channel as well.

Debate theatre

Diaboado Jacques Thiamobiga's chapter reports on one PDC tool developed to overcome a challenge found in some parts of Burkina Faso where women tend not to speak in public due to cultural restrictions (see 'Burkina Faso: When Farm Wives Take to the Stage', in Part IV). This difficult aspect of the communities' tradition became a major obstacle when actions to curb farmland erosion and desertification were initiated in the villages of Toukoro and Badara. Given the key role that women play in agriculture, the project was primarily aimed at working with peasant women. However, it soon became clear that another tradition that had not yet been discussed openly was hindering the full participation of women in these environmental efforts. Indeed, in this part of Burkina Faso, women are not entitled to landownership and are only granted the temporary right to grow food crops in some fields, which are usually the least productive. This makes it pointless for them to invest time and resources in upgrading a piece of land that will most likely be taken away from them once they become more productive, thus affecting the whole community. Since they were not allowed to speak up, they faced a big communication problem on top of all of the other problems.

The women of Badara were then reminded of a traditional ceremony where women are allowed to dress up as men and publicly talk about issues in a forthright manner. The women decided to exercise this right and to speak up in the form of debate-theatre performances.

The PDC team worked closely with the women in researching the messages to be presented, and helped them to prepare for their performances. When the play was ready, the women gave five performances: two in Badara, and one each in Toukoro, Tondogosso and Dou. Every performance was a big hit. It turned out to be fun for the women when they got to perform in their own village. They astonished the audience when they made their entrance on stage. Members of the audience were pleasantly surprised to see their mothers, sisters or wives dressed up as men. Many husbands recognized their wives through their gestures and their manner of speaking. In fact, each performance sparked a big celebration in the villages. The debate-theatre performances served, more importantly, as learning experiences for everyone: men, women and children. People exchanged viewpoints about village life; they discussed male–female relations and landownership; and they shared technical information for improving and preserving soil fertility. The unthinkable happened: the women succeeded in giving technical advice on soil fertility to the men. At the same time, the women found it an empowering experience during which they enhanced their own self-esteem and proved to the men that they were able to discuss their communities' development problems effectively in public and to propose useful solutions and courses of action.

Multiple methods along the Lisungwi, Mwanza and Mkulumadzi rivers

The case contributed by Meya Kalindekafe in Malawi made use of an interesting combination of participatory methods (see 'Participatory Research and Water Resource Management: Implementing the Communicative Catchment Approach in Malawi', in Part III). Meya's team began with the familiar research and development steps of literature review, training, surveys using semi-structured questionnaires, focus group discussions, key informant interviews and the Harvard Analytical Framework. It then proceeded to use more participatory methods involving members of the riverside communities in preparing resource maps and benefit–analysis charts. The team also went on transect walks with members of the communities, a method used frequently by PRA practitioners.

The work in Malawi is an example of the direction that many PDC teams will take, in the future, of carefully selecting tools and methods from across a number of disciplines for application within one initiative. Communication researchers have adopted multidisciplinary approaches after recognizing long ago that the field of communication is inherently multidisciplinary and not just narrowly confined within the boundaries of media tools and methods.

Efforts in the future

PDC is probably the youngest component in the relatively new field of communications. The experiments reported in this book, together with work attempted elsewhere, have proven that the assumptions behind PDC are sound. They also show that PDC approaches do help people to facilitate long-term development efforts on their own terms.

The recent advent of PDC has meant that many tools and methods remain to be fully developed. The complex group dynamics that PDC attempts to facilitate will require PDC practitioners to innovate these new tools and methods. The experiences in this book show that media-based methods are suitable for a good number of the processes that practitioners need to engage with. These experiences also demonstrate that many of the processes are interpersonal in nature or involve participatory group processes that seem best served by interpersonal or group communication techniques. This indicates the need for a range of interpersonal communication techniques and group facilitation methods to be developed or acquired by PDC practitioners. People-embodied methods are some of the most challenging to develop and to share; an effective facilitator of participatory group processes is easy to appreciate, but very difficult to emulate. Table 6.1 summarizes the availability or lack of various tools and methods for the different processes that PDC needs to be equipped with in order to support or facilitate.

Table 6.1 shows that many of the missing or emerging tools and methods are clustered around the two areas of creating knowledge and building communities. This reflects the novelty of PDC within these two areas, not previously highlighted by the communication field.

PDC probably has much to learn from the field of education in the area of creating knowledge. The possibilities of interdisciplinary efforts in this area are numerous and are potentially of mutual benefit to both the education and communications sectors.

The area of building community also stands to benefit from the experiences of work conducted in organizational communication; community organizing; cultural studies; ethnography; indigenous knowledge; studies of power structures; negotiation and mediation; and psychology.

Conclusions

Nora Cruz Quebral pointed out in her chapter at the opening of this book ('Participatory Development Communication: An Asian Perspective', Part II) that communications theories and models have not replaced each other, but instead coexist productively. This coexistence is due to the different roles that communication plays in human and community interactions. Each theory and model serves a particular set of interactions and sets out to accomplish a different goal. PDC's role in natural resource management and sustainable development seems most promising from the set of case studies published in this book. But it is not an easy role to play. It carries with it all the trials and tribulations of the performing arts, where actors must spend long years of arduous apprenticeship before they are ready to perform on stage. Even then, the challenge changes from one performance to another as the mood and composition of the people in the theatre alter. It is not cold science: people rest at the core of what is attempted.

Table 6.1 *Availability of tools and methods for participatory development communication processes*

Processes	Availability of tools and methods		
	Exists	Being developed	Largely absent
Effective communication			
Self-expression	√		
Listening and understanding		√	
Creating knowledge			
Sourcing information	√		
Tapping indigenous knowledge		√	
Processing and validating information		√	
Sharing knowledge			√
Building communities			
Building trust		√	
Managing conflict and competition			√
Forging partnerships		√	
Reinforcing self-identity			√
Reflecting on the past and present	√		
Visioning the future	√		
Affirming values			√
Adjusting values			√
Enabling transparency in decision-making			√
Sharing benefits		√	
Enabling action			
Identifying problems	√		
Evolving solutions	√		
Nurturing a sense of guardianship of the commons		√	
Managing expectations		√	
Taking stock and pooling resources	√		
Sourcing complementary resources	√		
Advocating to stakeholders	√		
Mobilizing for action	√		
Evaluating action	√		
Iterating and refining action	√		

PDC, however, is very different from the performing arts in many other aspects. The effort is not over between two predetermined curtain calls. PDC takes time, stretching more than 15 years in one case reported here. It also needs to be underpinned by a set of very clear principles and values.

While PDC is very much like other areas of communication in terms of the media and tools with which it works, it is radically different in its philosophy. So, while most areas of communication require their practitioners to articulate messages lucidly, PDC often requires us to be silent and to listen carefully.

References

Chambers, R. (1994) 'The origins and practice of participatory rural appraisal', *World Development,* vol 22, no 7, pp953–969

Green, L., Lallana, E. C., Shafiee, M. and Nain, Z. (2005) 'Social, political and cultural aspects of ICT: E-governance, popular participation and international politics', in Chin, S. Y. (ed) *Digital Review of Asia Pacific,* Southbound, Penang

Stonier, T. (1990) *Information and the Internal Structure of the Universe: An Exploration into Information Physics,* Springer, London

UNICEF (United Nations Children's Fund) Bangladesh (1993) *Visualization in Participatory Programmes: A Manual for Facilitators and Trainers Involved in Participatory Group Events,* UNICEF, Dhaka

Williamson, H. A. (1991) 'The Fogo Process: Development support communications in Canada and the developing world', in Casmir, F. L. (ed) *Communication in Development,* Ablex Publishing Corporation, Norwood

Selected Readings

C. V. Rajasunderam

This bibliography is divided into four parts:

1 introducing communication as a tool to improve community participation in participatory natural resource management (NRM) research;
2 communication for community participation in planning participatory NRM research;
3 communication for community participation in implementing and monitoring participatory NRM research;
4 communication and sharing knowledge.

I: Introducing communication as a tool to improve community participation in participatory natural resource management (NRM) research

Arnst, R. (1996) 'Participatory approaches to the research process', in Servaes, J. et al (eds) *Participatory Communication for Social Change*, Sage Publications, New Delhi, and Thousand Oaks, London, pp109–126
- *Content*: two major approaches to participatory communication are discussed in this contribution.
Belbase, S. (1994) 'Participatory communication in development: How can we achieve it', in White, A. et al (eds) *Participatory Communication: Working for Change and Development*, Sage Publications, New Delhi, and Thousand Oaks, London, pp446–461
- *Content:* describes the components of a participatory communication project with particular reference to a participatory development communication (PDC) project in Asia.
Cohen, S. I. (1996) 'Mobilizing communities for participation and empowerment', in Servaes, J. et al (eds) *Participatory Communication for Social Change*, Sage Publications, New Delhi, and Thousand Oaks, London, pp223–248
- *Content*: the focus is on participatory communication for community empowerment.
Diegues, S. A. C. (1992) 'Sustainable development and people's participation in wetland ecosystem conservation in Brazil: Two comparative case studies', in Ghai, D. and Vivian, M. (eds) *Grassroots Environmental Action – People's Participation in Sustainable Development*, Routledge, London and New York, pp141–158

- *Content*: this chapter documents the traditional resource management practices of two Brazilian communities and explores the ecological and social impacts of the disruption of those practices by state-supported schemes for increased levels of resource exploitation.

Egger, P. and Majeres, J. (1992) 'Local resource management and development: Strategic dimensions of people's participation', in Ghai, D. and Vivian, M. (eds) *Grassroots Environmental Action – People's Participation in Sustainable Development*, Routledge, London and New York, pp304–324

- *Content*: this chapter draws on a wide range of experiences to provide general lessons on participation in community-level sustainable development projects that are supported and/or initiated by external agencies. The example points to a number of strategic dimensions regarding the interaction between people, resource management and development.

Fraser, C. and Villet, J. (1994) 'Communication in practice', in Fraser, C. and Villet, J. (eds) *Communication: A Key to Human Development*, United Nations Food and Agriculture Organization, Rome, pp8–23

- *Content*: covers better planning and programme formulation; people's participation and communication mobilizing; changing lifestyles; improved training; rapid spread of information; effective management and coordination; generating the support of decision-makers.

Ingles, A. et al (1999) *The Participatory Process for Supporting Collaborative Management of Natural Resources: An Overview*, United Nations Food and Agriculture Organization, Rome

- *Content*: the overview to this book describes the extent and nature of participation in the collaborative management of natural resources. The processes and practical aspects of promoting collaborative management of natural resources are also discussed. This book is part of a new set of materials on participatory processes currently being developed by the Community Forestry Unit of the United Nations Food and Agriculture Organization (FAO).

Kennedy, T. (1989) 'Community animation – An open-ended process', *Communication for Community, WACC Media Development* 3/1989, World Association of Christian Communication (WACC), London, pp5–7

- *Content*: focuses on the variety of applications of the community animation approach as an alternative to the public hearing process, participatory research, organizational development, conflict resolution, urban renewal development and development communication.

Narayan, D. (1995) 'The contribution of people's participation: Evidence from 121 rural water supply projects', *Environmentally Sustainable Development Occasional Paper Series No 1*, World Bank, Washington, DC

- *Content*: this study examines efforts to induce participation as a means of creating effective rural water systems and building the local capacity to manage them.

Rahnema, M. (1992) 'Participation', in Sachs, W. (ed) *The Development Dictionary: A Guide to Knowledge as Power*, Zed Books, London and New Jersey, pp116–129

- *Content*: a useful contribution on the many dimensions of the participatory process in development with a focus on the philosophic premises underlying the concept.

Saik Yoon, C. (1996) 'Participatory communication for development', in Bessette, G. and Rajasunderam, C. V. (eds) *Participatory Communication for Development: A West African Agenda*, International Development Research Centre, Ottawa, Canada, and Southbound, Penang, pp37–61

- *Content:* a good introduction to the conceptual bases of PDC and the practical aspects of using this approach in field projects.

Servaes, J. and Arnst, R. (1993) 'First things first: Participatory communication for change', *Media Development Journal of the World Association of Christian Communication*, 2/1993, vol XL, pp44–47

- *Content:* focuses on a comparison of the 'mechanistic' and 'organic' models of development communication work.

Slocum, R. and Thomas-Slayter, B. (1995) 'Participation, empowerment and sustainable development', in Slocum, R. et al (eds) *Power, Process and Participation: Tools for Change*, Intermediate Technology Publications, London, pp3–8

- *Content:* provides the conceptual background for participatory research methodologies.

Thomas, P. (1994) 'Participatory development communication: Philosophical premises', in White, A. et al (eds) *Participatory Communication: Working for Change and Development*, Sage Publications, New Delhi, and Thousand Oaks, London, pp49–59

- *Content:* the focus of this contribution is on the potential and limitations of PDC.

Thomas-Slayter, B. (1995) 'A brief history of participatory methodologies', in Slocum, R. et al (eds) *Power, Process and Participation: Tools for Change*, Intermediate Technology Publications, London, pp9–16

- *Content:* a comparison of traditional research approaches with participatory research methods.

Vivian, J. M. (1992) 'Foundations for sustainable development: Participation, empowerment and local resource management', in Ghai, D. and Vivian, M. (eds) *Grassroots Environmental Action – People's Participation in Sustainable Development*, Routledge, London and New York, pp50–77

- *Content:* this contribution demonstrates in specific terms the importance of traditional resource management systems and locally based popular environmental initiatives.

White, S. A. (1994) 'The concept of participation: Transforming rhetoric to reality', in White, A. et al (eds) *Participatory Communication: Working for Change and Development*, Sage Publications, New Delhi, and Thousand Oaks, London, pp15–32

- *Content:* a good introduction to the basic principles of PDC.

II: Communication for community participation in planning participatory NRM research

The focus of this section is on the following themes:

- approaching a local community;
- collecting and sharing information;
- involving the community in identifying NRM problems and solutions.

Allen, W. and Kilvington, M. (2001) *ISKM (Integrated Systems for Knowledge Management)*, Landcare Research, Manaaki Whenua, www.landresearch.co.nz/research/social/iskm.asp

- *Content:* an outline of a participatory approach to environmental research and development initiatives. Managing the constructive involvement of stakeholders in NRM research is a skill. The Integrated Systems for Knowledge Management (ISKM) approach is designed to support such an ongoing process of constructive community dialogue. It provides clear communication pathways to support dialogue and collective action.

Allen, W. et al (2001) 'Using participatory and learning-based approaches for environmental management to help achieve constructive behaviour change, section 5: Concepts for managing participation in practice', Landcare Research, Manaaki Whenua, www.landcareresearch.co.nz/research/social/par_rep5.asp

- *Content:* this article examines specific mechanisms that collectively support an overall framework designed to facilitate behaviour change for environmental management. These mechanisms are social capital; empowering people and communities; levels of participation; managing a participatory process; stakeholder analysis; and participatory monitory and evaluation.

Bergdall, D. (1993a) *Methods for Active Participation – Experiences in Rural Development from East and Central Africa*, Oxford University Press, Nairobi, Kenya

- *Content:* this book comprises the first two parts of the final report on *The Methods for Active Participation Research and Development Project (MAP)*. The MAP project was implemented in Kenya, Tanzania and Zambia from 1988 to 1991.

Bergdall, D. (1993b) 'The MAP facilitator's handbook', in *Methods for Active Participation – Experiences in Rural Development from East and Central Africa*, Oxford University Press, Nairobi, Kenya, pp146–200

- *Content:* contains simple guidelines on the role of facilitators in catalysing community participation in development; techniques for enabling broad participation; factors for creating a participatory environment; and ensuring quality in the planning process.

Bessette, G. (2004) *Involving the Community: A Guide to Participatory Development Communication*, International Development Research Centre, Ottawa, Canada, and Southbound, Penang, Malaysia

- *Content:* this guide is intended for people working in research and development. It introduces PDC concepts, discusses the use of effective two-way communication approaches and presents a methodology to plan, develop and evaluate PDC projects.

Bidol, P. et al (eds) (1986a) 'Working definition of alternative environmental conflict management', in *Alternative Environmental Conflict Management Approaches: A Citizen's Manual*, School of Natural Resources, University of Michigan, Ann Arbor, Michigan, pp17–19

- *Content:* provides practical guidelines on conflict management approaches and related group problem-solving skills.

Bidol, P. et al (eds) (1986b) 'Overview of conflict management', in *Alternative Environmental Conflict Management Approaches: A Citizen's Manual*, School of Natural Resources, University of Michigan, Ann Arbor, Michigan, pp23–60

- *Content:* provides practical guidelines on conflict management approaches and related group problem-solving skills.

Bidol, P. et al (eds) (1986c) 'Effective citizen teams', in *Alternative Environmental Conflict Management Approaches: A Citizen's Manual*, School of Natural Resources, University of Michigan, Ann Arbor, Michigan, pp194–204

- *Content:* provides practical guidelines on conflict management approaches and related group problem-solving skills.

Chandrasekharan, D. (2000a) *Proceedings: Electronic Conference on Addressing Natural Resource Conflicts through Community Forestry*, United Nations Food and Agriculture Organization, Rome

- *Content:* this serves as a good introduction to the different types of conflicts in natural resource management.

Chandrasekharan, D. (2000b) 'Categorisation of conflicts – Annex F', in *Proceedings: Electronic Conference on Addressing Natural Resource Conflicts through Community Forestry*, United Nations Food and Agriculture Organization, Rome, pp137–143

Chandrasekharan, D. (2000c) 'Participatory approaches – Annex G', in *Proceedings: Electronic Conference on Addressing Natural Resource Conflicts through Community Forestry*, United Nations Food and Agriculture Organization, Rome, pp145–150

Chandrasekharan, D. (2000d) 'Definitions – Annex H', in *Proceedings: Electronic Conference on Addressing Natural Resource Conflicts through Community Forestry*, United Nations Food and Agriculture Organization, Rome, pp151–155

Cornwall, A. et al (1993) *Acknowledging Process: Challenges for Agricultural Research and Extension Methodology, Discussion Paper 3*, Institute of Development Studies, Brighton, UK, and International Institute for Environment and Development, London

- *Content:* this discussion paper locates agricultural research and extension practices within wider social processes. Six participatory approaches to research are discussed.

Feldstein, S. and Jiggins, J. (eds) (1994a) 'Participatory methodologies for analysing household activities, resources and benefits', in *Tools For The Field: Methodologies Handbook for Gender Analysis in Agriculture*, Kumarian Press and Intermediate Technology Publications, London, pp36–44

- *Content:* discusses tools for gender analysis in agriculture.

Feldstein, S. and Jiggins, J. (eds) (1994b) 'Workshops for gathering information', in *Tools For The Field: Methodologies Handbook for Gender Analysis in Agriculture*, Kumarian Press and Intermediate Technology Publications, London, pp55–61

- *Content:* strategies for collecting information on the role of women in agricultural initiatives at the community level.

Feldstein, S. and Jiggins, J. (eds) (1994c) 'Using focus groups with rural women', in *Tools For The Field: Methodologies Handbook for Gender Analysis in Agriculture*, Kumarian Press and Intermediate Technology Publications, London, pp62–65

- *Content:* discusses the focus group as a communication tool for catalysing community action.

Feldstein, S. and Jiggins, J. (eds) (1994d) 'Practical considerations for improving gender-based research', in *Tools For The Field: Methodologies Handbook for Gender Analysis in Agriculture,* Kumarian Press and Intermediate Technology Publications, London, pp236–238

- *Content:* useful guidelines for enhancing the effectiveness of research on gender issues in agriculture.

Feldstein, S. and Jiggins, J. (eds) (1994e) 'Women's agricultural production committees and the participative-research action approach', in *Tools For The Field: Methodologies Handbook for Gender Analysis in Agriculture*, Kumarian Press and Intermediate Technology Publications, London, pp239–243

- *Content:* an example of the learning process involved in the participative-research action approach.

Kaner, S. et al (1996a) 'Introduction', in *Facilitator's Guide to Participatory Decision-Making*, New Society Publishers, Philadelphia, pp XIII–XVI

- *Content:* an overview of the participatory decision-making process.

Kaner, S. et al (1996b) 'Participatory values', in *Facilitator's Guide to Participatory Decision-Making*, New Society Publishers, Philadelphia, pp23–29

- *Content*: discusses full participation; mutual understanding; inclusive solutions; and shared responsibility.

Kaner, S. et al (1996c) 'Introduction to the role of the facilitator', in *Facilitator's Guide to Participatory Decision-Making*, New Society Publishers, Philadelphia, pp31–37
- *Content*: outlines the expertise that supports a group to do its best thinking.

Kaner, S. et al (1996d) 'Facilitative listening skills', in *Facilitator's Guide to Participatory Decision-Making*, New Society Publishers, Philadelphia, pp41–53
- *Content*: presents techniques for honouring all points of view.

Kaner, S. et al (1996e) 'Facilitating open discussion', in *Facilitator's Guide to Participatory Decision-Making*, New Society Publishers, Philadelphia, pp55–67
- *Content*: promotes the free-flowing exchange of ideas.

Kaner, S. et al (1996f) 'Dealing with difficult dynamics', in *Facilitator's Guide to Participatory Decision-Making*, New Society Publishers, Philadelphia, pp113–167
- *Content*: discusses how a facilitator can intervene in a context of difficult group dynamics.

Kaner, S. et al (1996g) 'Building a shared framework of understanding', in *Facilitator's Guide to Participatory Decision-Making*, New Society Publishers, Philadelphia, pp169–182
- *Content*: describes the principles and tools that support groups struggling in the service of integration.

Mikkelsen, B. (1995) 'Participation: Concepts and methods', in *Methods for Development Work and Research, A Guide for Practitioners*, Sage Publications, New Delhi, and Thousand Oaks, London, pp61–82
- *Content*: presents useful guidelines on collecting development information by using participatory approaches.

Narayan, D. (1996a) 'Introduction', in *Towards Participatory Research*, World Bank Technical Paper, no 307, World Bank, Washington, DC, pp1–15
- *Content*: discusses why participatory processes in data collection are important.

Narayan, D. (1996b) 'What is Participatory Research?', in *Towards Participatory Research*, World Bank Technical Paper, no 307, World Bank, Washington, DC, pp17–31
- *Content*: discusses the roles of a participatory researcher, and how conventional and participatory research/data collection differ.

Narayan, D. (1996c) 'Defining the purpose of the study', in *Towards Participatory Research*, World Bank Technical Paper, no 307, World Bank, Washington, DC, pp33–45
- *Content*: presents methods for clarifying the research purpose and defining the scope of the study.

Narayan, D. (1996d) 'Choosing data collection methods', in *Towards Participatory Research*, World Bank Technical Paper, no 307, World Bank, Washington, DC, pp59–79
- *Content*: discusses matching methods with information needs.

Narayan, D. (1996e) 'Data analysis dissemination and use', in *Towards Participatory Research*, World Bank Technical Paper, no 307, World Bank, Washington, DC, pp129–138
- *Content*: methods for analysing data and sharing information are discussed in detail.

Norris, J. (1995) *An Introduction to Participatory Action Research Excerpted from the Guide to the Film* From the Field, PAR Trust, Calgary, Alberta
- *Content*: a lucid description of the participatory action research process.

Oltheten, T. M. P. (1995a) 'Major lessons learned from the case studies', in *Participatory Approaches to Planning for Community Forestry: Results and Lessons from Case Studies Conducted in Asia, Africa and Latin America – A Synthesis Report,* United Nations Food and Agriculture Organization, Rome, pp25–36
• *Content*: the focus of this publication is on participatory planning for community forestry as a learning process.
Oltheten, T.M.P. (1995b) 'Conclusions and recommendations', in *Participatory Approaches to Planning for Community Forestry: Results and Lessons from Case Studies Conducted in Asia, Africa and Latin America – A Synthesis Report,* United Nations Food and Agriculture Organization, Rome, pp37–43
Riano, P. (1991) 'Myths of the silenced: Women and grassroots communication', *Media Development,* 2/1991 vol XXXVIII, World Association of Christian Communication (WACC), London, pp20–22
• *Content*: this chapter focuses on the social functions and perceived roles of women in development.
Riano, P. (ed) (1994a) 'Women's participation in communication: Elements for a framework', in *Women in Grassroots Communication: Furthering Social Change,* Thousand Oaks, London, and Sage Publications, New Delhi, pp3–29
Riano, P. (ed) (1994b) 'Process video: Self-reference and social change', in *Women in Grassroots Communication: Furthering Social Change,* Thousand Oaks, London, and Sage Publications, New Delhi, pp131–148
Riano, P. (ed) (1994c) 'The WEDNET Initiative: A sharing experience between researchers and rural women', in *Women in Grassroots Communication: Furthering Social Change,* Thousand Oaks, London, and Sage Publications, New Delhi, pp221–234
• *Content*: focuses on the communicative roles of women at the grassroots level and on the use of participatory approaches for research on environmental issues. WEDNET stands for Women's Environment and Development Network.
Spring, A. (1996) 'Gender and environment: Some methods for extension specialists', in *Training for Agriculture,* United Nations Food and Agriculture Organization, Rome, pp104–122
• *Content*: the focus of this chapter is on the methods and techniques for analysing gender and natural resource management. The examples are taken from Africa, Asia and Latin America.
Srinivasan, L. (1990a) 'Community participation for development', in *Tools for Community Participation: A Manual for Training Trainers in Participatory Techniques,* Prowwess/UNDP, New York, pp15–19
• *Content*: a concise introduction to the concept of community participation.
Srinivasan, L. (1990b) 'Planning a participatory training programme', in *Tools for Community Participation: A Manual for Training Trainers in Participatory Techniques,* Prowwess/UNDP, New York, pp21–30
• *Content*: useful guidelines on planning a training programme in participatory techniques.
Srinivasan, L. (1990c) 'Designing the participatory workshop', in *Tools for Community Participation: A Manual for Training Trainers in Participatory Techniques,* Prowwess/ UNDP, New York, pp35–43
• *Content*: describes the sequence of steps involved in designing a workshop on tools for community participation.
Srinivasan, L. (1990d) 'Simple daily evaluation activities and techniques', in *Tools for Community Participation: A Manual for Training Trainers in Participatory Techniques,* Prowwess/UNDP, New York, pp45–50
• *Content*: outlines participatory methods of evaluating daily training activities.

III: Communication for community participation in implementing and monitoring participatory NRM research

This section focuses on the following themes:

- developing communication strategies;
- using communication tools;
- evaluating and documenting communication activities.

Allen, W. et al (2001) 'Building group capacity for environmental change', in *Using Participatory and Learning-Based Approaches for Environmental Management to Help Achieve Constructive Behaviour Change,* Landcare Research, Manaaki Whenua, www.landcareresearch.co.nz/research/social/par_rep6.asp
- *Content:* the focus of this article is on participatory techniques for working with groups and teams.
Berrigan, F. J. (1979a) 'The practice of community communication', in *Community Communication: The Role of Community Media in Development*, UNESCO, Paris, pp18–27
- *Content:* a discussion of the challenges involved in the practice of participatory communication at the community level.
Berrigan, F. J. (1979b) 'The community media methodology', in *Community Communication: The Role of Community Media in Development*, UNESCO, Paris, pp28–41
- *Content:* community media methodology is described in detail, with three examples of successful projects.
Braden, S. and Huong, T.T.T. (1998a) 'The role of video in participatory development', in *Video for Development*, Oxfam, Oxford, pp13–26
- *Content:* this casebook examines an experiment in the Ky Nam village, Viet Nam, where a team of non-governmental organization (NGO) workers from four countries were trained in the participatory use of video for community development. The team then worked with the villagers in Ky Nam to make videotapes about issues identified and researched by the community. The casebook gives an objective account of the workshop, showing how community-made video can be used locally for purposes of conflict resolution and advocacy. This particular chapter outlines some of the main concerns that lay behind the Oxfam initiative in Ky Nam and the key principles underlying the uses of participatory video.
Braden, S. and Huong, T. T. T. (1998b) 'Participatory research and analysis in Ky Nam', in *Video For Development*, Oxfam, Oxford, pp41–50
- *Content:* this chapter describes the ways in which the villagers used video to research and retrieve information about their needs and problems. It also focuses on the learning processes involving the villagers and the visiting team of NGO workers.
Carey, H. A. (1999a) 'Visual communication', in *Communication in Extension: A Teaching and Learning Guide*, United Nations Food and Agriculture Organization, Rome, pp53–66
- *Content:* guidelines for conducting a training/teaching session on designing visual aids.
Carey, H. A. (1999b) 'Social action processes', in *Communication in Extension: A Teaching and Learning Guide*, United Nations Food and Agriculture Organization, Rome, pp69–73

- *Content*: this chapter outlines a plan for conducting a training/teaching session on applying social action processes.

Carey, H. A. (1999c) 'Your communication/interaction traits', in *Communication in Extension: A Teaching and Learning Guide*, United Nations Food and Agriculture Organization, Rome, p116

- *Content*: this chapter presents questions that enable you to see yourself as others see you.

Carey, H. A. (1999d) 'Communication self-evaluation form', in *Communication in Extension: A Teaching and Learning Guide*, United Nations Food and Agriculture Organization, Rome, p117

- *Content*: a discussion of self-evaluation of communication skills.

Chambers, R. (2002) *Participatory Workshops: A Sourcebook of 21 Sets of Ideas and Activities*, Earthscan, London

- *Content:* this sourcebook of practical approaches and methods for participatory workshops draws on a rich variety of experiences and is a very useful book for facilitators and trainers.

CMN (Community Media Network) (1998a) 'What is community photography?', in *Tracking: Community Photography Report*, CMN, Dublin, p1

- *Content*: community photography as a tool for empowerment.

CMN (1998b) 'Some poor cousin?', in *Tracking: Community Photography Report*, CMN, Dublin, p2

- *Content*: current approaches to community photography.

CMN (1998c) 'The business of images', in *Tracking: Community Photography Report*, CMN, Dublin, p5

- *Content*: community photography as a tool for reflecting a community's life experiences.

Estrella, M. et al (eds) (2000a) 'Methodological issues in participatory monitoring and evaluation', in *Learning from Change: Issues and Experiences in Participatory Monitoring and Evaluation*, Intermediate Technology Publications, London, and International Development Research Centre, Ottawa, Canada, pp201–216

- *Content*: practical aspects and challenges of participatory monitoring and evaluation.

Estrella, M. et al (eds) (2000b) 'Laying the foundation: Capacity building for participatory monitoring and evaluation', in *Learning from Change: Issues and Experiences in Participatory Monitoring and Evaluation*, Intermediate Technology Publications, London, and International Development Research Centre, Ottawa, Canada, pp217–228

- *Content*: key elements of capacity-building for participatory monitoring and evaluation.

Estrella, M. et al (eds) (2000c) 'Learning to change by learning from change: Going to scale with participatory monitoring and evaluation', in *Learning from Change: Issues and Experiences in Participatory Monitoring and Evaluation*, Intermediate Technology Publications, London, and International Development Research Centre, Ottawa, Canada, pp229–243

- *Content*: focuses on participatory monitoring and evaluation as a learning process involving many different stakeholders. This chapter also explores some of the social and political dimensions of participatory monitoring and evaluation, especially in relation to scaling up.

FAO (United Nations Food and Agriculture Organization) (1990) 'The communication system for Proderith II', in *Towards Putting Farmers in Control: A Second Case Study of the Rural Communication System for Development in Mexico's Tropical Wetlands*, FAO, Rome, pp11–15

- *Content*: focuses on the network approach and the various types of media used in this approach to rural development.

FAO (2000a) 'An alternative to literacy: Is it possible for community video and radio to play this role? A small experiment by the Deccan Development Society in Hyderabad, India', *Forests, Trees and People Newsletter,* no 40/41, December 1999/January 2000, FAO, Rome, pp9–13

- *Content*: these articles are devoted to the use of popular communication tools within a participatory process.

FAO (2000b) 'The development of participatory media in southern Tanzania', *Forests, Trees and People Newsletter,* no 40/41, December 1999/January 2000, FAO, Rome, pp14–18

FAO (2000c) 'Participatory video and PRA: Acknowledging the politics of empowerment', *Forests, Trees and People Newsletter,* no 40/41, December 1999/January 2000, FAO, Rome, pp21–23

FAO (2000d) 'What is all this song and dance about?', *Forests, Trees and People Newsletter,* no 40/41, December 1999/January 2000, FAO, Rome, pp24–25

FAO (2000e) 'TV Favela, a voice of the people: Experience with local TV in environmental work in Brazil', *Forests, Trees and People Newsletter* no 40/41, December 1999/January 2000, FAO, Rome, pp26–29

FAO (2000f) 'On becoming visible: The role of video in community strategies to take part in the discussion about the future of the forest', *Forests, Trees and People Newsletter,* no 40/41, December 1999/January 2000, FAO, Rome, pp30–34

FAO (2000g) 'Questions and answers about participatory video', *Forests, Trees and People Newsletter,* no 40/41, December 1999/January 2000, FAO, Rome, pp35–44

FAO (2000h) 'Community forestry radio over Nepal', *Forests, Trees and People Newsletter,* no 40/41, December 1999/January 2000, FAO, Rome, pp45–47

FAO (2000i) 'Networking for dialogue and action', *Forests, Trees and People Newsletter,* no 40/41, December 1999/January 2000, FAO, Rome, pp54–58

Feek, W. and Morry, C. (2003) *Communication and Natural Resource Management: Theory-Experience*, FAO, Rome, and The Communication Initiative, Victoria, British Columbia, www.fao.org/documents/show_cdr.asp?url_file=DOCREP/005/Y4737E/Y4737E00.HTM

- *Content*: this book has been written as a tool for people involved or interested in communication and natural resource management (CNRM). It presents short case studies, reflections and exercises that guide the reader through a self-learning process about CNRM practice and theory.

de Fossard, E. and Kulakow, A. (n.d. a) 'Overview of the steps of the planning process for development communication', in *A Planning Guide for Development Communication,* Foundation For International Training, Markham, Ontario, and Academy of Educational Development, Washington, DC, pp9–23

- *Content*: useful guidelines on the communication planning process. This chapter includes brief descriptions of 17 major steps involved in planning a development communication project.

de Fossard, E. and Kulakow, A. (n.d. b) 'Media glossary', in *A Planning Guide for Development Communication,* Foundation For International Training, Markham, Ontario, and Academy of Educational Development, Washington, DC, pp86–93

- *Content*: summary of the potential and limitations of different media used in development communication.

de Fossard, E. and Kulakow, A. (n.d. c) 'Guidelines for sample testing', in *A Planning Guide for Development Communication,* Foundation For International Training, Markham, Ontario, and Academy of Educational Development, Washington, DC, pp139–149

- *Content*: outline of a plan for pre-testing development communication materials.

de Fossard, E. and Kulakow, A. (n.d. d) 'Evaluation guide', in *A Planning Guide for Development Communication,* Foundation For International Training, Markham, Ontario, and Academy of Educational Development, Washington, DC, pp149–158

- *Content*: describes the instruments for formative and summative evaluation of development communication projects; it also provides useful guidelines on devising the evaluation questionnaire.

Guijt, I. (1998a) *Participatory Monitoring and Impact Assessment of Sustainable Agricultural Initiatives*, SARL Discussion Paper no 1, International Institute for Environment and Development, London

- *Content*: this discussion paper is a practical methodological introduction to setting up a participatory monitoring process for sustainable agricultural initiatives.

Guijt, I. (1998b) 'Definitions', in *Participatory Monitoring and Impact Assessment of Sustainable Agricultural Initiatives*, SARL Discussion Paper no 1, International Institute for Environment and Development, London, pp12–21

- *Content*: clarifies key concepts.

Guijt, I. (1998c) 'The key steps', in *Participatory Monitoring and Impact Assessment of Sustainable Agricultural Initiatives*, SARL Discussion Paper no 1, International Institute for Environment and Development, London, pp22–26

Guijt, I. (1998d) 'Indicators', in *Participatory Monitoring and Impact Assessment of Sustainable Agricultural Initiatives*, SARL Discussion Paper no 1, International Institute for Environment and Development, London, pp27–29

Guijt, I. (1998e) 'Annex I', in *Participatory Monitoring and Impact Assessment of Sustainable Agricultural Initiatives*, SARL Discussion Paper no 1, International Institute for Environment and Development, London, pp53–110

Harford, N. and Baird, N. (1997a) 'Introduction: Guidelines for making visual aids', in *How to Make and Use Visual Aids*, Volunteer Services Overseas, Heinemann Educational Publishers, Oxford, pp5–19

- *Content*: this chapter covers planning for making visual aids according to the type of activity; it also provides simple guidelines on pre-testing and evaluating audiovisual materials.

Harford, N. and Baird, N. (1997b) 'Re-usable visual aids which carry information', in *How to Make and Use Visual Aids*, Volunteer Services Overseas, Heinemann Educational Publishers, Oxford, pp20–43

- *Content*: this chapter discusses objects that carry information; these objects can be used over and over again, conveying different information each time.

Harford, N. and Baird, N. (1997c) 'Visual aids which display information', in *How to Make and Use Visual Aids*, Volunteer Services Overseas, Heinemann Educational Publishers, Oxford, pp44–66

- *Content*: in this chapter, the reader is introduced to displays of information, which are made more effective by using simple graphic design techniques.

IIED (International Institute for Environment and Development) (2000a) 'Keys to unleash mapping's good magic', in *PLA Notes, Participatory Learning and Action 39 – Special Issue on Popular Communications*, IIED, London, pp10–13

- *Content*: this article describes the eight steps involved in community-based mapping and the key questions that should be answered at each stage.

IIED (2000b) 'A participatory GIS for community forestry user groups in Nepal: Putting people before the technology', in *PLA Notes, Participatory Learning and Action 39 – Special Issue on Popular Communications*, IIED, London, pp14–18

- *Content*: this article explores some of the benefits and concerns of using geographical information systems (GIS) as a participatory tool.

IIED (2000c) 'Flying to reach the sun', in *PLA Notes, Participatory Learning and Action 39 – Special Issue on Popular Communications*, IIED, London, pp59–60
- *Content*: an interview with Anna Blackman of Photo Voice, a UK-based non-profit organization that works on participatory documentary and photography projects around the world.

Kennedy, T. W. (1982) 'Beyond advocacy: A facilitative approach to public participation', *Journal of the University Film and Video Association*, University Film and Video Association, vol XXXIV, 3/summer 1982, pp??[Q206]
- *Content*: the focus of this article is on the use of film and video as a catalyst for community discussions of development issues.

Linney, B. (1995a) 'Approaches to communication', in *Pictures, People and Power: People-centred Visual Aids for Development*, Macmillan, London, pp5–8
- *Content*: reflections on the authoritarian approach (one-way communication) and the people-centred approach (two-way communication).

Linney, B. (1995b) 'Making and using people-centred pictures', in *Pictures, People and Power: People-centred Visual Aids for Development*, Macmillan, London, pp87–131
- *Content*: guidelines on making and using a range of visual aids for people-centred communication.

Linney, B. (1995c) 'Pictures and empowerment', in *Pictures, People and Power: People-centred Visual Aids for Development*, Macmillan, London, pp173–188
- *Content*: the role of people-centred pictures in empowering individuals, groups and communities.

Narayan, D. (1993a) 'Introduction', in *Participatory Evaluation Tools for Managing Change in Water and Sanitation*, World Bank Technical Paper no 207, World Bank, Washington, DC, pp1–7
- *Content*: the role of evaluation in community-managed projects.

Narayan, D. (1993b) 'What is participatory evaluation', in *Participatory Evaluation Tools for Managing Change in Water and Sanitation*, World Bank Technical Paper no 207, World Bank, Washington, DC, pp9–19
- *Content*: focuses on problem-solving orientation, generating knowledge and involving users in analytical data.

Narayan, D. (1993c) 'Measuring sustainability', in *Participatory Evaluation Tools for Managing Change in Water and Sanitation*, World Bank Technical Paper no 207, World Bank, Washington, DC, pp27–30; pp43–61
- *Content*: issues covered in this article include human capacity development; managing abilities; local institutional capacity; knowledge and skills; systems for learning and problem-solving; and supportive leadership.

Nelson, N. and Wright, S. (eds) (1995) 'Theatre for development: Listening to the community', in *Power and Participatory Development, Theory and Practice*, Intermediate Technology Publications, London, pp61–71
- *Content*: an interesting case study of a community environment project in east Mali, which used a drama unit to tune into the mood and views expressed by the community.

Oakley, P. et al (1991a) 'Emerging methodologies of participation', in *Projects With People: The Practice of Participation in Rural Development*, International Labour Organization, Geneva, pp205–238
- *Content*: discusses the methodological tools used to promote participation in development.

Oakley, P. et al (1991b) 'Evaluating participation', in *Projects With People: The Practice of Participation in Rural Development*, International Labour Organization, Geneva, pp239–268

- *Content*: describes the conceptual challenges of evaluating the process of community participation.

Pretty, J. N. et al (1995a) 'You the trainer and facilitator', in *A Training Guide for Participatory Learning and Action*, International Institute for Environment and Development, London, pp13–38
- *Content*: focuses on the principal roles, skills and techniques that the trainer-facilitator should consider before undertaking participatory training activities.

Pretty, J. N. et al (1995b) 'Group dynamics and team-building', in *A Training Guide for Participatory Learning and Action*, International Institute for Environment and Development, London, pp39–53
- *Content*: the main features of managing group dynamics and building inter-disciplinary teams that are essential in practising participatory research and development.

Pretty, J. N. et al (1995c) 'Principles of participatory learning and action', in *A Training Guide for Participatory Learning and Action*, International Institute for Environment and Development, London, pp54–71
- *Content*: a summary of the core principles of participatory learning and action.

Pretty, J. N. et al (1995d) 'Training in participatory methods in the workshop', in *A Training Guide for Participatory Learning and Action*, International Institute for Environment and Development, London, pp72–89
- *Content*: describes the process of workshop training in three groups studying participatory methods.

Pretty, J. N. et al (1995e) 'The challenges of training in the field', in *A Training Guide for Participatory Learning and Action*, International Institute for Environment and Development, London, pp90–110
- *Content*: the complexities of training in a real world setting and how to deal with them.

Pretty, J. N. et al (1995f) 'Organising workshops for training, orientation and exposure', in *A Training Guide for Participatory Learning and Action*, International Institute for Environment and Development, London, pp11–129
- *Content*: the conditions necessary for preparing a training course on participatory methods.

Ramirez, R. and Quarry, W. (2004) *Communication for Development: A Medium for Innovation in Natural Resource Management*, IDRC, Ottawa, Canada, and FAO, Rome
- *Content*: this report presents, through stories and examples, the experiences of many people and projects worldwide where communication methods and approaches have been applied to address natural resource management.

Ray, H. E. (1984) 'Guidelines for planning and implementation of communication strategies', in *Incorporating Communication Strategies into Technology Transfer Programmes for Agricultural Development*, Academy of Educational Development, Washington, DC, pp85–133
- *Content*: a set of useful guidelines for planning and implementing effective communication programmes as integral components of agricultural projects.

Richardson, D. and Rajasunderam, C. V. (2000) 'Training community animators as participatory communication for development practitioners', in Richardson, D. and Paisley, L. (eds) *The First Mile of Connectivity*, www.fao.org/docrep/x0295e/x0295e17.htm
- *Content*: this contribution explores key questions and issues related to the training of animators/facilitators in PDC processes.

Servaes, J. et al (eds) (1996) 'Fitting projects to people or people to projects', in *Participatory Communication for Social Change*, Sage Publications, New Delhi, and Thousand Oaks, London, pp249–265
- *Content*: describes nine steps to develop an effective communication strategy for projects.
Shaw, J. and Robertson C. (1997a) 'Background, approaches and benefits', in *Participatory Video: A Practical Guide to Using Video Creatively in Group Development Work*, Routledge, London and New York, pp7–27
- *Content*: the benefits of participatory video (community-building; critical awareness; consciousness raising; empowerment; and self-reliance) are discussed in terms of group purpose, community purpose and individual purpose.
Shaw, J. and Robertson C. (1997b) 'Applications and project outcomes', in *Participatory Video: A Practical Guide to Using Video Creatively in Group Development Work*, Routledge, London and New York, pp166–191
- *Content*: a comprehensive guide to using video in a participatory way in group development work.
Slocum, R. et al (1996a) 'Community drama', in *Power, Process and Participation: Tools for Change*, Intermediate Technology Publications, London, pp72–74
- *Content*: this selection offers a description of field tested communication and participatory research tools for securing people's participation in development projects.
Slocum, R. et al (1996b) 'Conflict resolution', in *Power, Process and Participation: Tools for Change*, Intermediate Technology Publications, London, pp75–87
Slocum, R. et al (1996c) 'Focus groups', in *Power, Process and Participation: Tools for Change*, Intermediate Technology Publications, London, pp95–99
Slocum, R. et al (1996d) 'Network formation', in *Power, Process and Participation: Tools for Change*, Intermediate Technology Publications, London, pp155–158
Slocum, R. et al (1996e) 'Photography', in *Power, Process and Participation: Tools for Change*, Intermediate Technology Publications, London, pp167–171
Slocum, R. et al (1996f) 'Social network mapping', in *Power, Process and Participation: Tools for Change*, Intermediate Technology Publications, London, pp186–190
Slocum, R. et al (1996g) 'Study trips', in *Power, Process and Participation: Tools for Change*, Intermediate Technology Publications, London, pp191–193
Slocum, R. et al (1996h) 'Transects', in *Power, Process and Participation: Tools for Change*, Intermediate Technology Publications, London, pp198–204
Slocum, R. et al (1996i) 'Video', in *Power, Process and Participation: Tools for Change*, Intermediate Technology Publications, London, pp205–213
White, S. A. et al (1994a) 'Facilitating communication within rural and marginal communities: A model for development support', in *Participatory Communication: Working for Change and Development*, Sage Publications, London and New Delhi, pp329–341
- *Content*: focuses on strategies to mobilize community groups for development action.
White, S. A. et al (1994b) 'Participatory message making with video: Revelations from studies in India and the USA', in *Participatory Communication: Working for Change and Development*, Sage Publications, London and New Delhi, pp359–383
- *Content*: the two studies described in this contribution focus on the process of using the participatory message development model.
White, S. A. (ed) (1999) *Facilitating Participation: Releasing the Power of Grassroots Communication*, Sage Publications, New Delhi, and Thousand Oaks, London

- *Content*: this book is divided into three broad sections: 'The art of activation', 'The art of technique' and 'The art of community building'. Part 1 includes a presentation of the important concept of the catalyst communicator. Part 2 explores significant issues related to the competencies of facilitators. Part 3 contains three case studies that focus on the link between participatory communication and community-building.

Windahl, S. et al (1992a) 'The nature of communication planning', in *Using Communication Theory: An Introduction to Planned Communication,* Sage Publications, London and New Delhi, pp19–29
- *Content*: a broad conceptualization of the communication planning process.

Windahl, S. et al (1992b) 'Categorisation of basic strategies', in *Using Communication Theory: An Introduction to Planned Communication,* Sage Publications, London and New Delhi, pp39–49
- *Content*: this chapter discusses basic types of communication strategies.

IV: Communication and sharing knowledge

This section focuses on the following themes:

- facilitating the identification and sharing of local knowledge;
- planning dissemination of research results to different stakeholders;
- facilitating extension to other communities.

Allen, W. K. and Harmsworth, M. G. (2001) *The Role of Social Capital in Collaborative Learning,* Landcare Research, Manaaki Whenua, www.landcareresearch.co.nz/research/social/social_capital.asp
- *Content*: 'social capital' (a term used in development and organizational learning literature) can be thought of as the networks of communication and cooperation that facilitate collaborative learning. The role of social capital in fostering the social networks and information exchange required for collective action is the focus of this contribution.

Bolliger, E. et al (1992a) 'Extension methods', in *Agricultural Extension: Guidelines for Extension Workers in Rural Areas – Part B. List of Questions,* Swiss Centre for Development Cooperation in Technology and Management (SKAT), ??[Q239], Chapter 4
- *Content*: guidelines on agricultural extension methods – for example, demonstrations, field trips and competitions.

Bolliger, E. et al (1992b) 'Extension accessories', in *Agricultural Extension: Guidelines for Extension Workers in Rural Areas – Part B. List of Questions,* Swiss Centre for Development Cooperation in Technology and Management (SKAT), ??[Q241], Chapter 5
- *Content*: guidelines on the use of extension accessories – for example, technical leaflets, plots, displays and rural newspapers.

ETG Netherlands (1997) 'Spreading and consolidating the PTD process', in *Developing Technology With Farmers: A Trainer's Guide for Participatory Learning,* Zed Books, London and New York, and ETG, The Netherlands, pp191–213
- *Content*: participatory technology development (PTD) in agriculture is a process of purposeful and creative interaction between local people and outside facilitators in order to develop more sustainable farming systems. The focus of this chapter

is on the last phase of the PTD process, in which the outcome of experimental activities by and with farmers is spread to other farmers.

FAO (United Nations Food and Agriculture Organization) (1996a) 'Integrating science and traditional knowledge to achieve sustainable development in Morocco', in *Training for Agriculture and Rural Development 1995–1996,* FAO, Rome, pp79–85

- *Content*: describes two Moroccan projects in order to demonstrate the importance of integrating traditional and scientific knowledge.

FAO (1996b) 'Group-based extension programmes in Java to strengthen natural resource conservation activities', in *Training for Agriculture and Rural Development 1995–1996,* FAO, Rome, pp123–133

- *Content*: describes how agricultural extension programmes in Indonesia are carried out through a group-based participatory approach. This chapter also records Indonesia's experience with field schools – an effective method for promoting farmers' learning and participation in agriculture.

Grenier, L. (1998a) 'What about indigenous knowledge?', in *Working with Indigenous Knowledge: A Guide for Researchers*, International Development Research Centre, Ottawa, Canada, pp1–11

- *Content*: describes some characteristics of indigenous knowledge and its contribution to sustainable development.

Grenier, L. (1998b) 'Developing a research framework', in *Working with Indigenous Knowledge: A Guide for Researchers*, International Development Research Centre, Ottawa, Canada, pp31–55

- *Content*: presents considerations for those who are developing frameworks for indigenous knowledge research.

Grenier, L. (1998c) 'Data collection', in *Working with Indigenous Knowledge: A Guide for Researchers*, International Development Research Centre, Ottawa, Canada, pp57–62

- *Content*: presents details of 31 field techniques for data collection.

Grenier, L. (1998d) 'Assessing, validating and experimenting with IK', in *Working with Indigenous Knowledge: A Guide for Researchers*, International Development Research Centre, Ottawa, Canada, pp71–86

- *Content*: assesses the products of IK research in terms of sustainability and looks at developing IK through validation and experimentation.

Grenier, L. (1998e) 'Sample guidelines', in *Working with Indigenous Knowledge: A Guide for Researchers*, International Development Research Centre, Ottawa, Canada, pp87–99

- *Content*: presents three sets of formal procedural guidelines for conducting IK research. The guidelines can be adapted for other situations.

Haverkort, B. et al (eds) (1991) 'Farmers' experiments and participatory technology development', in *Joining Farmers' Experiments: Experiences in Participatory Technology Development*, Intermediate Technology Publications, London, pp3–16

- *Content*: outlines the six steps in participatory technology development:
  1 how to get started;
  2 looking for things to do;
  3 designing the experiment;
  4 trying out;
  5 sharing results with others; and
  6 sustaining and consolidating the process of participatory technology development.

Inglis, J. T. (ed) (1993a) 'Traditional ecological knowledge in perspective', in *Traditional Ecological Knowledge: Concepts and Cases*, International Programme on Traditional Ecological Knowledge and International Development Research Centre, Ottawa, Canada, pp1–6
- *Content*: outlines the reasons for preserving traditional ecological knowledge (TEK), with a focus on the practical uses of TEK.
Inglis, J. T. (ed) (1993b) 'Transmission of traditional ecological knowledge', in *Traditional Ecological Knowledge: Concepts and Cases*, International Programme on Traditional Ecological Knowledge and International Development Research Centre, Ottawa, Canada, pp17–30
- *Content*: addresses the key issue of how ecological knowledge is transmitted from one generation to the next.
Inglis, J. T. (ed) (1993c) 'Integrating traditional ecological knowledge and management with environmental impact assessment', in *Traditional Ecological Knowledge: Concepts and Cases*, International Programme on Traditional Ecological Knowledge and International Development Research Centre, Ottawa, Canada, pp33–39
- *Content*: provides perspectives on the use of traditional ecological knowledge in environmental assessment.
Inglis, J. T. (ed) (1993d) 'African indigenous knowledge and its relevance to sustainable development', in *Traditional Ecological Knowledge: Concepts and Cases*, International Programme on Traditional Ecological Knowledge and International Development Research Centre, Ottawa, Canada, pp55–62
- *Content*: discusses the relevance of African indigenous knowledge to environment and development issues.
Johnson, M. (ed) (1992a) 'Research on traditional environmental knowledge: Its development and its role', in *LORE: Capturing Traditional Environmental Knowledge*, Dene Cultural Institute and International Development Research Centre, Ottawa, Canada, pp3–22
- *Content*: focuses on the problems of integrating traditional environmental knowledge and Western science.
Johnson, M. (ed) (1992b) 'Documenting and applying traditional environmental knowledge in northern Thailand', in *LORE: Capturing Traditional Environmental Knowledge*, Dene Cultural Institute and International Development Research Centre, Ottawa, Canada, pp164–173
- *Content*: describes the efforts of the Mountain People's Culture and Development Education Programme to document and apply the TEK of the highlanders of northern Thailand.
Narayan, D. (1996) 'Data analysis, dissemination and use', in *Towards Participatory Research*, World Bank Technical Paper no 307, World Bank, Washington, DC, pp129–138
- *Content*: explores techniques for data analysis; dissemination and utilization of research results; media options for communicating results; and follow-up activities.
Starkey, P. (1999a) 'Network types and network benefits', in *Networking for Development*, International Forum for Rural Transport and Development (IFRTD), London, pp14–20
- *Content*: examines the benefits of networking in development work; some network models are also discussed.
Starkey, P. (1999b) 'General guidelines for networking', in *Networking for Development*, International Forum for Rural Transport and Development (IFRTD), London, pp31–46

- *Content*: sets out concise, practical guidelines for an effective network operation, with examples from a range of development networks.

Van Veldhuizen, L. et al (ed) (1997) 'Kuturaya: Participatory research, innovation and extension', in *Farmers' Research in Practice: Lessons From The Field*, Intermediate Technology Publications, London, pp153–173

- *Content:* a case study of a research project on conservation tillage initiated by the Zimbabwe Agricultural Extension Service, AGRITEX Kuturaya, which emphasized learning and improving through experimentation. Farmers found a renewed confidence in their own knowledge and abilities as a result of their experimentation.

Warner, K. (1991) 'Local technical knowledge, shifting cultivation and natural resource management', in *Shifting Cultivators: Local Technical Knowledge and Natural Resource Management in the Humid Tropics*, United Nations Food and Agriculture Organization, Rome, pp1–10

- *Content:* examines shifting cultivation as a natural resource management strategy for the tropics, with a focus on utilizing local technical knowledge.

Warren, M. et al (eds) (1995a) 'Indigenous communication and indigenous knowledge', in *The Cultural Dimension of Development: Indigenous Knowledge Systems*, Intermediate Technology Publications, London, pp112–123

- *Content*: describes the interface between indigenous knowledge and indigenous communication. A framework for studying this interface is also outlined.

Warren, M. et al (eds) (1995b) 'Farmer know-how and communication for technology transfer: CTTA in Niger', in *The Cultural Dimension of Development: Indigenous Knowledge Systems*, Intermediate Technology Publications, London, pp323–332

- *Content*: describes the model of communication elaborated and applied by the Communication for Technology Transfer in Agriculture (CTTA) programme in its research on farmer innovation and communication in Niger.

Warren, M. et al (eds) (1995c) 'Indigenous soil and water conservation in Djenne, Mali', in *The Cultural Dimension of Development: Indigenous Knowledge Systems*, Intermediate Technology Publications, London, pp371–384

- *Content*: describes indigenous soil and water conservation practices and techniques in sub-Saharan Africa in the circle of Djenne, central Mali.

Index